1,000,000 Books
are available to read at

Forgotten Books

www.ForgottenBooks.com

Read online
Download PDF
Purchase in print

ISBN 978-1-330-13416-0
PIBN 10034157

This book is a reproduction of an important historical work. Forgotten Books uses state-of-the-art technology to digitally reconstruct the work, preserving the original format whilst repairing imperfections present in the aged copy. In rare cases, an imperfection in the original, such as a blemish or missing page, may be replicated in our edition. We do, however, repair the vast majority of imperfections successfully; any imperfections that remain are intentionally left to preserve the state of such historical works.

Forgotten Books is a registered trademark of FB &c Ltd.
Copyright © 2018 FB &c Ltd.
FB &c Ltd, Dalton House, 60 Windsor Avenue, London, SW19 2RR.
Company number 08720141. Registered in England and Wales.

For support please visit www.forgottenbooks.com

1 MONTH OF FREE READING

at
www.ForgottenBooks.com

By purchasing this book you are eligible for one month membership to ForgottenBooks.com, giving you unlimited access to our entire collection of over 1,000,000 titles via our web site and mobile apps.

To claim your free month visit:
www.forgottenbooks.com/free34157

* Offer is valid for 45 days from date of purchase. Terms and conditions apply.

English
Français
Deutsche
Italiano
Español
Português

www.forgottenbooks.com

Mythology Photography **Fiction**
Fishing Christianity **Art** Cooking
Essays Buddhism Freemasonry
Medicine **Biology** Music **Ancient Egypt** Evolution Carpentry Physics
Dance Geology **Mathematics** Fitness
Shakespeare **Folklore** Yoga Marketing
Confidence Immortality Biographies
Poetry **Psychology** Witchcraft
Electronics Chemistry History **Law**
Accounting **Philosophy** Anthropology
Alchemy Drama Quantum Mechanics
Atheism Sexual Health **Ancient History**
Entrepreneurship Languages Sport
Paleontology Needlework Islam
Metaphysics Investment Archaeology
Parenting Statistics Criminology
Motivational

First Year Algebra Scales

Henry Gustave Hotz, Ph.D.

Teachers College, Columbia University
Contributions to Education, No. 90

Published by
𝕿eachers College, Columbia University
New York City
1918

Copyright, 1918, by HENRY GUSTAVE HOTZ

EDUC.
LIBRARY

ACKNOWLEDGMENT

Contributions to this study have come from so many sources that it is impossible for me to adequately acknowledge all of them here. Hundreds of teachers, principals, and superintendents have willingly coöperated in securing the sixteen thousand papers upon which the results of this study are based. To them I am also indebted for many valuable suggestions and criticisms. More particularly, however, I desire here to express to Professor George D. Strayer and to Professor Edward L. Thorndike my very grateful appreciation of their continued assistance and sympathetic supervision at each stage of this study.

H. G. H.

CONTENTS

I. Introduction 1
   Problem and Method
   Recent Attempts to Develop Standards in Algebra

II. Origin and Use of First Year Algebra Scales . . 3
   1. History of Their Derivation
   2. Directions for Giving the Tests
   3. Instructions for Scoring and Tabulating Results
   4. Determination of a Class Score
   5. Tentative Standard Scores

III. Derivation of the Equation and Formula Scale . 42
   1. Arrangement of Material
   2. Probable Error (P. E.) Taken as the Unit of Measure
   3. Scaling the Exercises in the Equation and Formula Test for Each Group
   4. Measurement of Distances between the Medians of Successive Groups
   5. Establishing the Zero Point
   6. Location of Each Exercise upon a Linear Scale

IV. Tables of Crude Data from Which Other Scales Were Derived 71
   1. Addition and Subtraction Scale
   2. Multiplication and Division Scale
   3. Problem Scale
   4. Graph Scale

V. Appendix—Administration and Use of First Year Algebra Scales 81

FIRST YEAR ALGEBRA SCALES

I

INTRODUCTION

The application of scientific measurement to schoolroom products has created a demand for objective standards. Indeed, progress in the evaluation of educational results is at present largely conditioned upon the successful establishment of reliable standards and units for measurement, and it may also be added that nowhere are these instruments of measurement more badly needed than in the field of secondary education.

The purpose of this study is to derive a series of scales for the measurement of achievement in first year algebra. The 'number of times correctly solved' is used as the basis for the statistical calculations in the development of these scales, and the method employed coincides most closely with that used by Doctor Trabue[1] in his *Completion-Test Language Scales* and by Doctor Woody[2] in his *Measurements of Some Achievements in Arithmetic.*

In designing these scales the writer felt that the main business of the work in first year algebra was to teach students how to solve the typical algebra problems through the use of simple algebraic symbols and devices. This makes the equation, the formula, the graph, and to a lesser degree the proportion, the important instruments to be used in algebra. Hence, in the construction of these scales the major emphasis was placed upon these phases of quantitative thinking. For the work in the fundamentals, scales in addition and subtraction and in multiplication and division were devised.

Each scale includes only such exercises or problems as are quite universally taught in a course in first year algebra. An attempt was also made to include as large a variety of exercises as could possibly be included, and the writer firmly believes that no essential algebraic process has been ignored. As usual in scale construction,

---

[1] Trabue, Marion Rex: *Completion-Test Language Scales,* 1916.
[2] Woody, Clifford: *Measurements of Some Achievements in Arithmetic,* 1916.

the exercises begin with very easy ones, which can readily be solved by nearly all students, but become increasingly more difficult so that the last ones in each series can be solved by only a relatively small number of the students who try them.

Numerous attempts to develop standards in algebra have already been made. Professor Thorndike[3] started the work in 1914. He selected a list of twenty-five problems and then had two hundred teachers of mathematics rate them in order of difficulty. Since then standard tests in algebra have been devised by Doctor Rugg,[4] by W. S. Monroe,[5] by H. G. Childs,[6] and by C. E. Stromquist.[7] A Scale for Testing Ability in Algebra has also been constructed by W. H. Coleman.[8]

[3] Thorndike, E. L.: "An Experiment in Grading Problems in Algebra" in *Mathematics Teacher*, 6:123, 1914.
[4] Rugg, H. O.: "Standardized Tests in First Year Algebra" in *School Review*, 25:113, 196, 346, 1917.
[5] Monroe, W. S.: "A Test of Attainment of First Year High School Students in Algebra" in *School Review*, 23:159, 1915.
[6] Childs, H. G.: "Measurement of Achievement in Algebra"; in Third Conference on Educational Measurement, in *Bulletin of Extension Division*, Indiana University, 1916, vol. II, no. 6, p. 171.
[7] University of Wyoming, Laramie, Wyoming.
[8] Superintendent of Schools, Bertrand, Nebraska.

II

ORIGIN AND USE OF THE FIRST YEAR ALGEBRA SCALES

1. HISTORY OF THEIR DERIVATION

The preliminary work on the scales, as shown in their final form on the next few pages, was begun in the fall of 1916. The first set was drawn up in accordance with the principles previously announced and included tests in:

1. Addition and Subtraction
2. Multiplication and Division
3. Equation and Formula
4. Problems
5. Graphs

These tests were given during the months of December, January, and February in various high schools of New York, New Jersey, and Connecticut. In all, 3,605 test papers were secured for this preliminary series. These were distributed as follows: The addition and subtraction test was given to 829 students, the multiplication and division test to 818 students, the equation and formula test to 800 students, the problem test to 895 students, and the graph test to 263 students.

From the results of this preliminary work it was evident that many of the exercises selected were useless, or, at best, of doubtful value as test material; some of them were not clearly stated, others ran very unevenly as seen by the fact that they were often solved more frequently by classes having little training in algebra than by more mature classes, and still others tended to group themselves too much about the same point upon a scale of relative difficulty. It was, therefore, necessary to discard many of the original exercises. Those remaining were then arranged in order of difficulty, as determined by the preliminary results, and wherever large gaps appeared between two successive exercises an attempt was made to interpose new ones. In these final revisions, as well as in the original construction of the tests, only those exercises and problems most commonly found in the recent textbooks were included. In the selection of all material due consideration was also given to suggestions of competent teachers and specialists in mathematics.

The revised tests were given during the months of April, May, and June, 1917. They were given to students who had studied algebra all the way from two to ten months. Practically all of those tested were in the ninth grade or the so-called first year of the high school.

Classes were tested in eighty-four high schools located in the states of Massachusetts, Connecticut, Rhode Island, New York, New Jersey, Ohio, Wisconsin, Missouri, Oklahoma, Colorado, and Washington. These schools varied in size all the way from the small rural high school to the large cosmopolitan high school. The addition and subtraction exercises were given to 3,186 students, the multiplication and division exercises to 3,065 students, the equation and formula exercises to 3,047 students, the problem set to 2,670 students, and the graphs to 1,137 students.

By far the greater number of the tests were given by the writer himself.[1] The method adopted in administering the tests is given in detail in the following section. In order to be absolutely certain that the exact difficulty of each exercise or problem be accurately determined, it would have been desirable to allow each pupil all the time he wished to use on each test. This was, however, not always possible on account of the rigid daily schedule upon which many schools were operating, and it was often necessary to call time at the end of forty or forty-five minutes. Evidence collected by a system of checks, however, seems to indicate that the time allotment was ample [2] and that the results were not materially affected by such limitation.

[1] Nearly all of those given in the East were given by the writer. Those given in the West and Middle West were given by men who were known personally and who could be trusted to follow instructions faithfully.

[2] In ninety-six out of two hundred of the tests submitted to nine months' students, exercises No. 19 and No. 25 in the equation and formula test were interchanged. They were then submitted to classes on a 'fifty-fifty' basis. The results showed that exercise No. 25 was solved correctly about three times as often when it came last in the list, while No. 19, on the other hand, was solved more frequently when it came nineteenth on the list. Similar checks were employed in each of the other tests, with the exception of the graph test, and similar results were obtained.

Origin and Use 5

SERIES A

FIRST YEAR ALGEBRA SCALES*

Write your name here.............................Age.......

When did you begin to study algebra? Month..........Year......

ADDITION AND SUBTRACTION

Carefully perform the operations as indicated.

(2) 1. $4r + 3r + 2r =$ 3. $12b + 6b - 3b =$

(7) 5. $7x - x + 6 - 4 =$ (8) 9. $(4r - 5t) + (s - 3r) =$

  10. $8c - (-6 + 3c) =$ 12. $5x - [4x - (3x - 1)] =$

13. $\dfrac{3c}{4} - \dfrac{3c}{8} =$ (17) 15. $\dfrac{1}{a-x} - \dfrac{3x}{a^2 - x^2} =$

(15) 18. $3 - \dfrac{3 - 2x}{4} - 2x =$ 22. $\dfrac{3 - 2x}{(x-1)^3} + \dfrac{x+1}{(x-1)^2} - \dfrac{1}{(x-1)} =$

(24) 23. $\sqrt{20} + \sqrt{45} + \sqrt{\tfrac{1}{5}} =$ (23) 24. $\dfrac{a}{a-2} - \dfrac{a-2}{a+2} + \dfrac{3}{4-a^2} =$

---

*These scales are printed separately on large sheets 8½" x 11". Sufficient space is left between the exercises so the pupils can work directly upon these sheets.

A number appearing in parenthesis before one of the above exercises or problems indicates that it occupied a different position in the test series from that now occupied in the final scales. In the statistical tables of this study the numbering of the exercises corresponds to that used in the test series.

First Year Algebra Scales

SERIES A
FIRST YEAR ALGEBRA SCALES

Write your name here............................Age......

When did you begin to study algebra? Month.........Year......

MULTIPLICATION AND DIVISION

Carefully perform the operations as indicated. Reduce all answers to their *simplest* forms.

1. $3 \cdot 7y =$

2. $\dfrac{12n}{4} =$

3. $2a \cdot 4ab^2 =$

7. $4x \cdot (-3xy^3) =$

(8) 9. $\dfrac{18m^2n - 27mn^2}{9mn} =$

11. $(2a^2 + 7a - 9)(5a - 1) =$

16. $(-3xy^3)^4 =$

(18) 17. $\dfrac{c^4 - d^4}{(c-d)^2} \cdot \dfrac{c-d}{c^2 + d^2} =$

(22) 20. $\dfrac{p^2 + 4p - 45}{p^3 + 2p + 4} \cdot \dfrac{p^3 - 8}{p^2 - 81} \cdot \dfrac{1}{3pr - 15r} =$

(20) 21. $\dfrac{x^3 + 27}{x^2 + x - 12} \div \dfrac{3x + 9}{x + 4} =$

(21) 22. $64^{\frac{2}{3}} \times 27^{\frac{1}{3}} =$

23. $\dfrac{3\sqrt{6a}}{2a\sqrt{18}} \cdot \sqrt{12a}$

Origin and Use

SERIES A

FIRST YEAR ALGEBRA SCALES

Write your name here ....Age........

When did you begin to study algebra? Month..........Year......

EQUATION AND FORMULA

Solve the following equations and formulae:

1. $2x = 4$.

(3) 2. $7m = 3m + 12$.

4. $5a + 5 = 61 - 3a$.

(5) 6. $10 - 11z = 4 - 8z$.

(7) 8. $c - 2(3 - 4c) = 12$.

(9) 11. The area of a triangle $= \tfrac{1}{2}bh$, in which
    $b =$ length of the base
    and $h =$ height of the triangle.

How many square feet are there in the area of a triangle whose base is 10 feet, and whose height is 8 feet?

(16) 14. $3m + 7n = 34$
        $7m + 8n = 46$

(17) 18. $\dfrac{x+3}{x-2} = \dfrac{x+5}{x-4}$.

19. $p^2 - 5p = 50$.

(21) 23. $\dfrac{6x-2}{x+3} - 3 = \dfrac{3x^2+13}{x^2-9}$.

24. $S = \tfrac{1}{2}gt^2$, solve for t.

25. $\sqrt{x^2 - 1} - x = -1$.

SERIES A

FIRST YEAR ALGEBRA SCALES

Write your name here.................................Age.......
When did you begin to study algebra? Month.........Year......

PROBLEMS

Do not work out the answer to the problem—merely *indicate* the answer or *state* the equation in each case.

1. If one coat cost x dollars, how much will 3 coats cost?
    Answer........................

(3) 2. A man is m years old; how old was he r years ago?
    Answer........................

4. A gold watch is worth ten times as much as a silver watch, and both together are worth $132. How much is each worth?
    Equation........................

5. The distance from Chicago to New York by rail is 980 miles. If a train runs v miles an hour, what is the time required for the run?
    Answer........................

7. The total number of circus tickets sold was 836. The number of tickets sold to adults was 136 less than twice the number of children's tickets. How many were sold of each?
    Equation........................

8. A rectangular box is d inches deep, w inches wide, and contains r cubic inches. What is its length?
    Answer........................

9. The area of a square is equal to that of a rectangle. The base of the rectangle is 12 feet longer and its altitude 4 feet shorter than the side of the square. Find the dimensions of both figures.
    Equation........................

Origin and Use 9

12. A train leaves a station and travels at the rate of 40 miles an hour. Two hours later a second train leaves the same station and travels in the same direction at the rate of 55 miles an hour. Where will the second train pass the first?

    Equation.................................

(14) 13. A merchant has two kinds of tea, one kind costing 50 cents and the other 65 cents per pound. How many pounds of each must be mixed together to produce a mixture of 20 pounds that shall cost 60 cents per pound?

    Equation.................................

(13) 14. An open box is made from a square piece of tin by cutting out a 5 inch square from each corner and turning up the sides. How large is the original square, if the box contains 180 cubic inches?

    Equation.................................

SERIES A

FIRST YEAR ALGEBRA SCALES

Write your name here.................................Age.......

When did you begin to study algebra? Month..........Year......

GRAPHS

1. The following diagram represents the length of certain rivers:

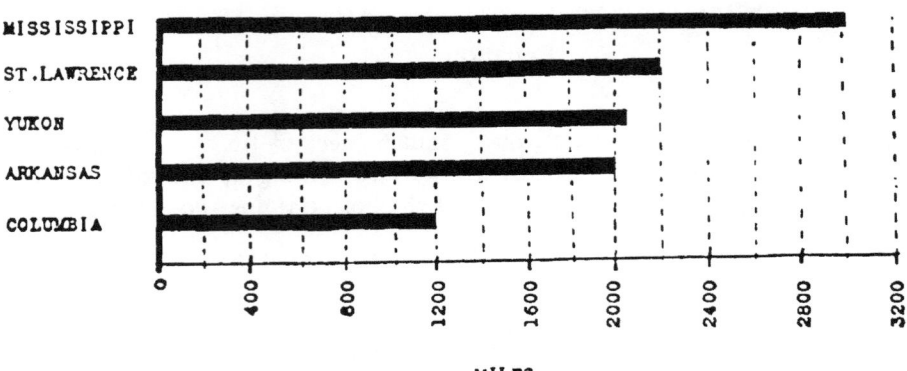

How many miles long is the Arkansas river as represented in this diagram?...

3. This graph indicates the population of a certain town for a period of years:

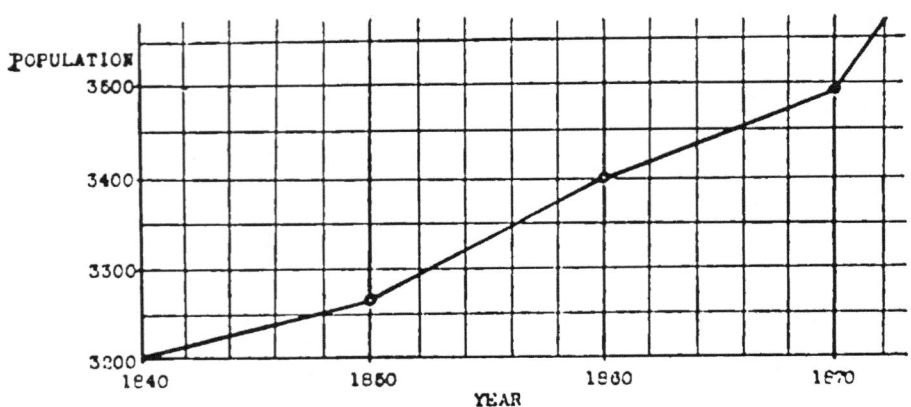

What was the approximate population of the town in 1856?..............

Origin and Use

4. This table gives the interest at 6 per cent on $100 for a period of years:

Years	Interest in dollars
1	6
2	12
3	18
4	24
5	30
6	36

Locate in the adjacent diagram the points corresponding to the above data.
Then connect these points by a line.

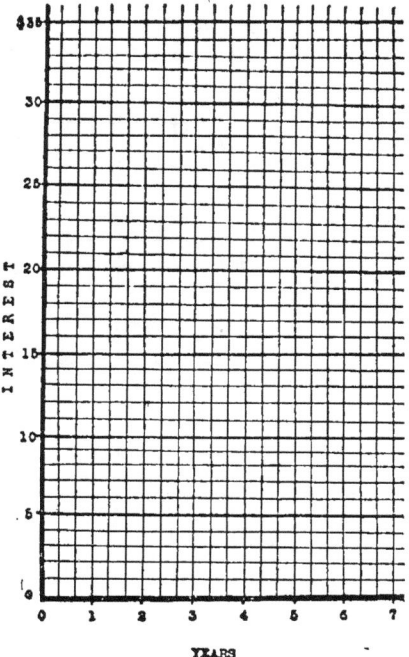

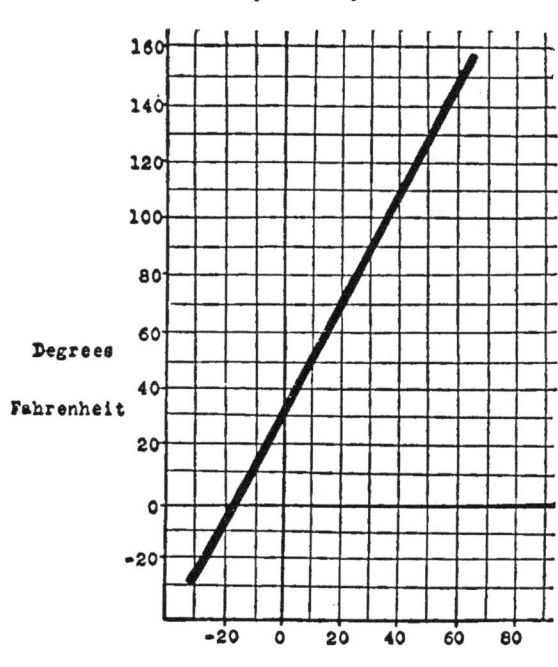

(6) 5. The graph opposite is used to convert degrees of temperature from the Fahrenheit scale (F) to the Centigrade scale (C), and from the Centigrade scale to the Fahrenheit scale.

When it is $+20°$ on the F. scale, what is the temperature in degrees on the C. scale? .....
....

(7) 6. This graph represents the distance passed over by a man walking at the rate of 3 miles an hour for a number of hours.

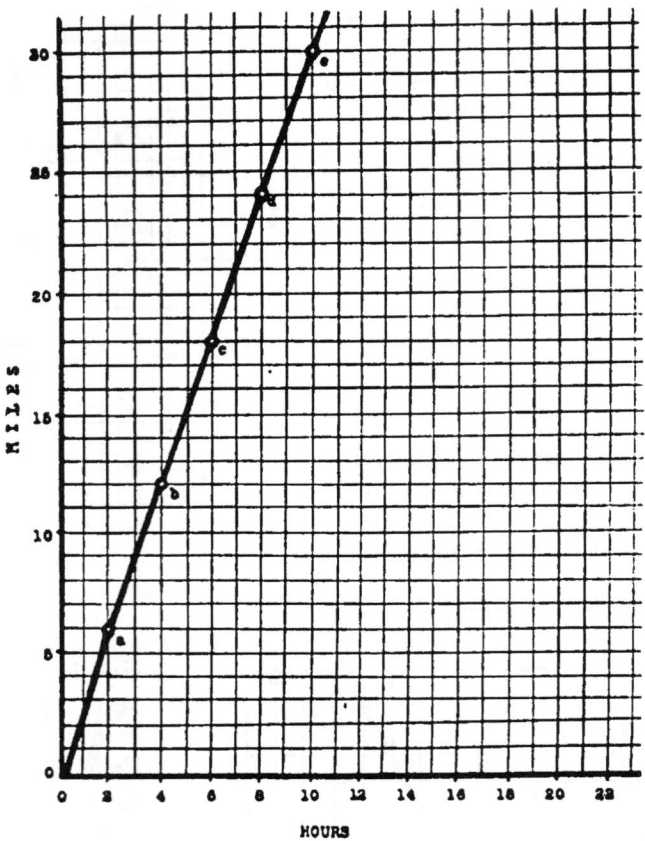

By the conditions of the problem it may be said that the number of miles travelled equals three times the number of hours, or

$$m = 3h$$

That is—

if $h = 2$, then $m = 6$;
if $h = 4$, then $m = 12$;
and if $h = 6$, then $m = 18$, etc.

By locating these points, we have a, b, c, d, etc., of the graph.

In the same diagram draw a graph in which

$$m = 2h$$

Origin and Use 13

(9) 7. Find three pairs of values for x and y in the following equation and then draw the graph of
$$x + y = 5.$$

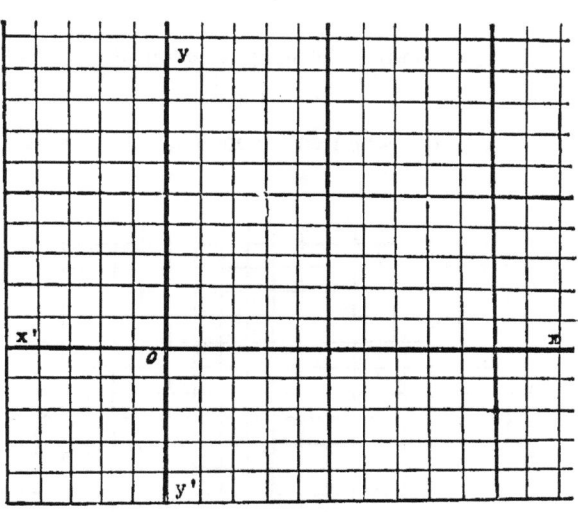

1 Space = 1 Unit

10. Plot the following equations and find the values of x and y at the point of intersection of the graphs.
$$x + 4y = 11.$$
$$2x - y = 4.$$

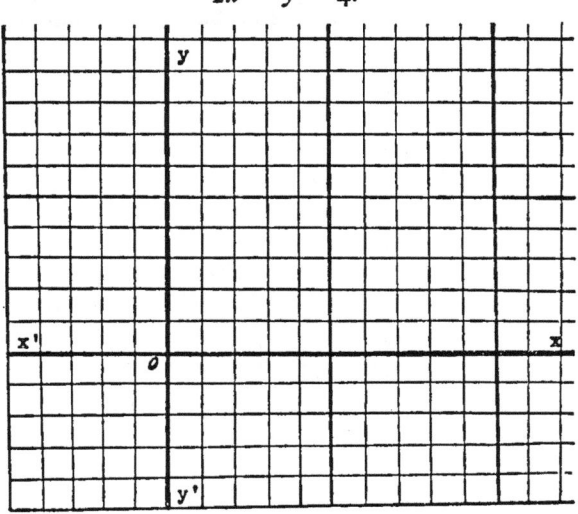

1 Space = 1 Unit

11. A boy begins work with a weekly wage of $9 and receives an increase of 25 cents every week. Another boy starts with a weekly wage of only $6 but receives an increase of 50 cents every week.

Draw a graph which shows the wage of each at the beginning of every week for 15 weeks.

According to this graph when will their wages be the same?............

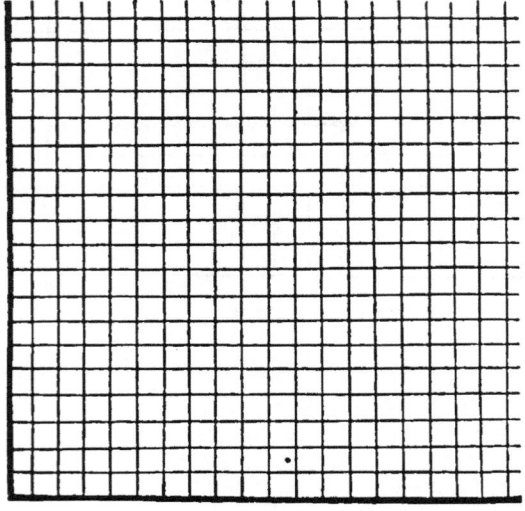

Origin and Use 15

SERIES B

FIRST YEAR ALGEBRA SCALES

Write your name here........................Age.......
When did you begin to study algebra? Month.........Year......

ADDITION AND SUBTRACTION

Carefully perform the operations as indicated.

(2) 1. $4r + 3r + 2r =$ (1) 2. $2x + 3x =$

3. $12b + 6b - 3b =$ 4. $2c + \frac{1}{2}c =$

(7) 5. $7x - x + 6 - 4 =$ (5) 6. $3a - 4b + 5a - 2b =$

(6) 7. $5m + (-4m) =$ (9) 8. $20x - (10x + 5x) =$

(8) 9. $(4r - 5t) + (s - 3r) =$ 10. $8c - (-6 + 3c) =$

11. $3a^2 - 3b - (2a^2 + 3b - 4) =$ 12. $5x - [4x - (3x - 1)] =$

13. $\dfrac{3c}{4} - \dfrac{3c}{8} =$ 14. $\dfrac{3x - 2}{3} + \dfrac{x + 4}{6} =$

(17) 15. $\dfrac{1}{a - x} - \dfrac{3x}{a^2 - x^2} =$ 16. $\dfrac{r}{r + z} + \dfrac{r}{r - z} =$

(18) 17. $\dfrac{5a + 1}{6a} - \dfrac{3a - 2}{2a} =$ (15) 18. $3 - \dfrac{3 - 2x}{4} - 2x =$

19. $\dfrac{10x + 3y}{2x^2y} - \dfrac{3x + 5y}{xy^2} =$

(21) 20. $\dfrac{1}{a + 1} - \dfrac{a}{a^2 - a + 1} - \dfrac{a - 4}{a^3 + 1} =$

(20) 21. $\dfrac{2}{x^2 - 5x + 6} - \dfrac{15}{x^2 + 2x - 15} =$

22. $\dfrac{3 - 2x}{(x - 1)^3} + \dfrac{x + 1}{(x - 1)^2} - \dfrac{1}{(x - 1)} =$

(24) 23. $\sqrt{20} + \sqrt{45} + \sqrt{\tfrac{1}{5}} =$

(23) 24. $\dfrac{a}{a - 2} - \dfrac{a - 2}{a + 2} + \dfrac{3}{4 - a^2} =$

SERIES B
FIRST YEAR ALGEBRA SCALES

Write your name here.................................Age.......
When did you begin to study algebra? Month..........Year......

MULTIPLICATION AND DIVISION

Carefully perform the operations as indicated. Reduce all answers to their *simplest* forms.

1. $3 \cdot 7y =$

2. $\dfrac{12n}{4} =$

3. $2a \cdot 4ab^2 =$

4. $6c^3 \div 2c^2 =$

5. $\tfrac{2}{3}$ of $9m =$

6. $\dfrac{-8a^2b}{4a^2} =$

7. $4x \cdot (-3xy^3) =$

(9) 8. $a^3 \cdot (-3a) \cdot (-2a) =$

(8) 9. $\dfrac{18m^2n - 27mn^2}{9mn} =$

(12) 10. $\dfrac{4x^4}{5} \div 2x^2 =$

11. $(2a^2 + 7a - 9)(5a - 1) =$

(13) 12. $\dfrac{n^4 + 7n^2 - 30}{n^2 - 3} =$

(14) 13. $\dfrac{7a}{15} \div \dfrac{7a^2}{20} =$

(10) 14. $\dfrac{-12x^2y^2 \cdot (x - 2)}{-3x^2y^2} =$

15. $\dfrac{m + n}{a} \cdot \dfrac{b}{m^2 - n^2} =$

16. $(-3xy^3)^4 =$

(18) 17. $\dfrac{c^4 - d^4}{(c - d)^2} \cdot \dfrac{c - d}{c^2 + d^2} =$

(17) 18. $3x^{\frac{1}{2}} \cdot 4x^{\frac{3}{2}} =$

19. $\dfrac{a^2 + \tfrac{3}{2}a - 1}{a + 2} =$

(22) 20. $\dfrac{p^2 + 4p - 45}{p^2 + 2p + 4} \cdot \dfrac{p^3 - 8}{p^2 - 81} \cdot \dfrac{1}{3pr - 15r} =$

(20) 21. $\dfrac{x^3 + 27}{x^2 + x - 12} \div \dfrac{3x + 9}{x + 4} =$

(21) 22. $64^{\frac{2}{3}} \times 27^{\frac{1}{3}} =$

23. $\dfrac{3}{2a}\sqrt{\dfrac{6a}{18}} \cdot \sqrt{12a}$

SERIES B

FIRST YEAR ALGEBRA SCALES

Write your name here.................................Age....

When did you begin to study algebra? Month.........Year....

EQUATION AND FORMULA

Solve the following equations and formulae·

    1. $2x = 4$. (3) 2. $7m = 3m + 12$.

(2) 3. $3x + 3 = 9$. 4. $5a + 5 = 61 - 3a$.

(6) 5. $7n - 12 - 3n + 4 = 0$. (5) 6. $10 - 11z = 4 - 8z$.

(8) 7. $\frac{2}{3}z = 6$. (7) 8. $c - 2(3 - 4c) = 12$.

(10) 9. $\frac{1}{2}x + \frac{1}{4}x = 3$. (11) 10. $\dfrac{2x}{3} = \dfrac{5}{8}$.

(9) 11. The area of a triangle = ½ bh, in which
                            b = length of the base
                  and h = height of the triangle.

How many square feet are there in the area of a triangle whose base is 10 feet, and whose height is 8 feet?

(13) 12. $\dfrac{y}{3} = \dfrac{5}{2} - \dfrac{y}{4}$. (12) 13. $\frac{1}{4}(x + 5) = 5$.

(16) 14. $\begin{array}{c}3m + 7n = 34\\ 7m + 8n = 46\end{array}$ (14) 15. $\dfrac{4}{3 - x} = \dfrac{2}{1 + x}$.

(15) 16. The area of a circle = πr^2, in which
                         r = radius of the circle
                  and $\pi = 3\frac{1}{7}$.

Find the area in square feet of circle whose radius in 7 feet.

(18) 17. In the formula $RM = El$, find the value of M.

(17) 18. $\dfrac{x + 3}{x - 2} = \dfrac{x + 5}{x - 4}$. 19. $p^2 - 5p = 50$.

18   *First Year Algebra Scales*

20. $\dfrac{2}{x^2 + 4x + 3} = \dfrac{3}{x^2 + 3x + 2}$.   (22) 21. $\dfrac{1}{x} + \dfrac{2}{y} = 1$

$\dfrac{4}{x} - \dfrac{4}{y} = 1$

(23) 22.   F = temperature in Fahrenheit degrees
C = temperature in Centigrade degrees
and $F = \dfrac{9C}{5} + 32°$.
Solve for C when F = 70°.

(21) 23. $\dfrac{6x - 2}{x + 3} - 3 = \dfrac{3x^2 + 13}{x^2 - 9}$.

24. $S = \tfrac{1}{2}gt^2$, solve for t.

25. $\sqrt{x^2 - 1} - x = -1$.

SERIES B

FIRST YEAR ALGEBRA SCALES

Write your name here........................Age.......
When did you begin to study algebra? Month.........Year......

PROBLEMS

Do not work out the answer to the problem—merely *indicate* the answer or *state* the equation in each case.

1. If one coat cost x dollars, how much will 3 coats cost?
Answer.....................

(3) 2. A man is m years old; how old was he r years ago?
Answer.....................

(2) 3. A boy has a marbles and buys b more; how many has he then?
Answer.....................

4. A gold watch is worth ten times as much as a silver watch, and both together are worth $132. How much is each worth?
Equation......................

Origin and Use

5. The distance from Chicago to New York by rail is 980 miles. If a train runs v miles an hour, what is the time required for the run?

    Answer.........................

6. The width of a basket ball court is 20 feet less than its length. The perimeter of the court (distance around) is 240 feet. Find the dimensions.

    Equation.........................

7. The total number of circus tickets sold was 836. The number of tickets sold to adults was 136 less than the number of children's tickets. How many were sold of each?

    Equation.........................

8. A rectangular box is d inches deep, w inches wide, and contains r cubic inches. What is its length?

    Answer.........................

9. The area of a square is equal to that of a rectangle. The base of the rectangle is 12 feet longer and its altitude 4 feet shorter than the side of the square. Find the dimensions of both figures.

    Equation.........................

10. A tower casts a shadow of 20 feet. A man, 5 feet 9 inches high, who is near at the same time, casts a shadow of 2 feet 6 inches. Find the height of the tower.

    Proportion.........................

11. Five thousand dollars is invested in two banks, part in one at 3 per cent and the rest in the other at 4 per cent. The annual income from the two investments is \$172. How much is each investment?

    Equation.........................

12. A train leaves a station and travels at the rate of 40 miles an hour. Two hours later a second train leaves the same station and travels in the same direction at the rate of 55 miles an hour. Where will the second train pass the first?

    Equation.........................

13. A merchant has two kinds of tea, one kind costing 50 cents and the other 65 cents per pound. How many pounds of each must be mixed together to produce a mixture of 20 pounds that shall cost 60 cents per pound?

    Equation.........................

14. An open box is made from a square piece of tin by cutting out a 5 inch square from each corner and turning up the sides. How large is the original square, if the box contains 180 cubic inches?

Equation..................................

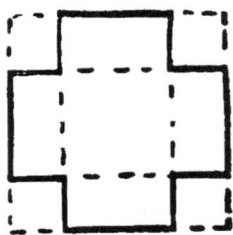

SERIES B

FIRST YEAR ALGEBRA SCALES

Write your name here.............................Age......

When did you begin to study algebra? Month..........Year......

GRAPHS

1. The following diagram represents the length of certain rivers:

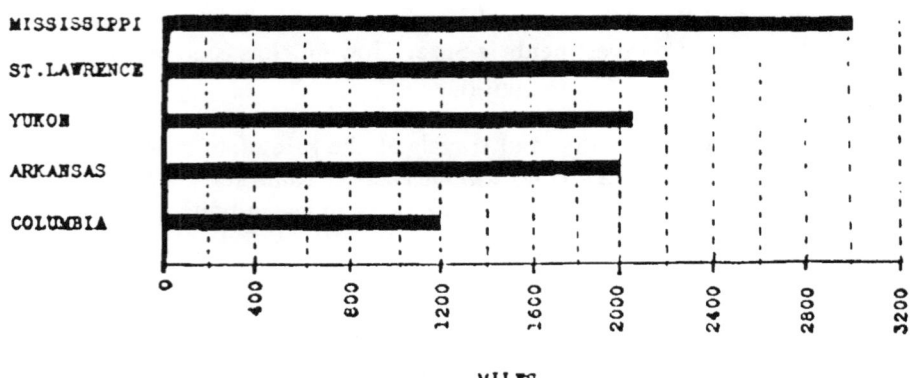

How many miles long is the Arkansas river as represented in this diagram?..............

2. The following graph represents the temperature at various hours of a certain day:

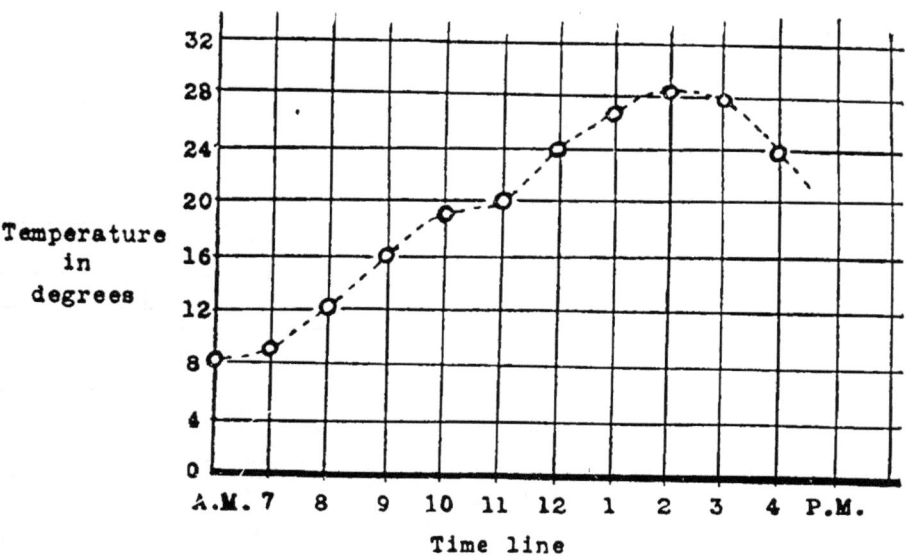

How many degrees was it at twelve o'clock?.....................

3. This graph indicates the population of a certain town for a period of years:

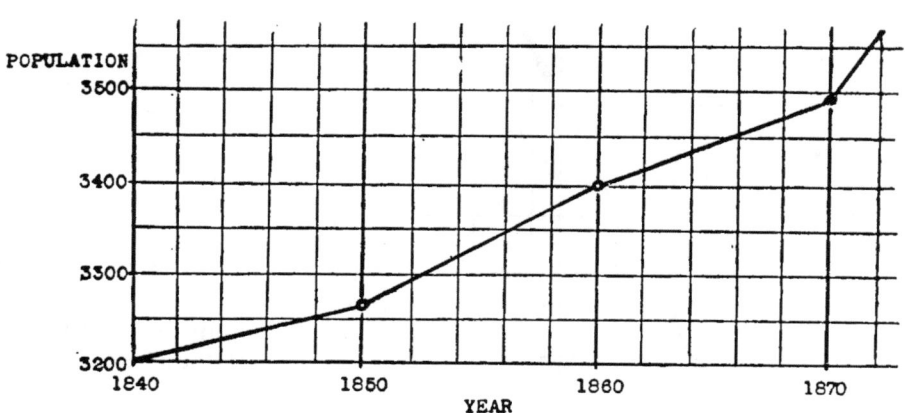

What was the approximate population of the town in 1856?........

4. This table gives the interest at 6 per cent on $100 for a period of years:

Years	Interest in dollars
1	6
2	12
3	18
4	24
5	30
6	36

Locate in the adjacent diagram the points corresponding to the above data.
Then connect these points by a line.

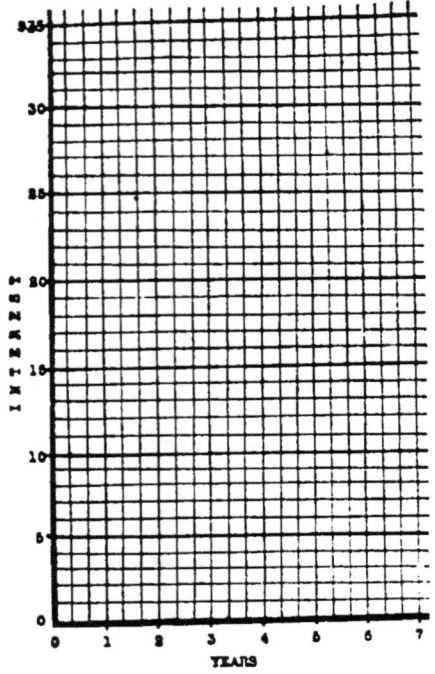

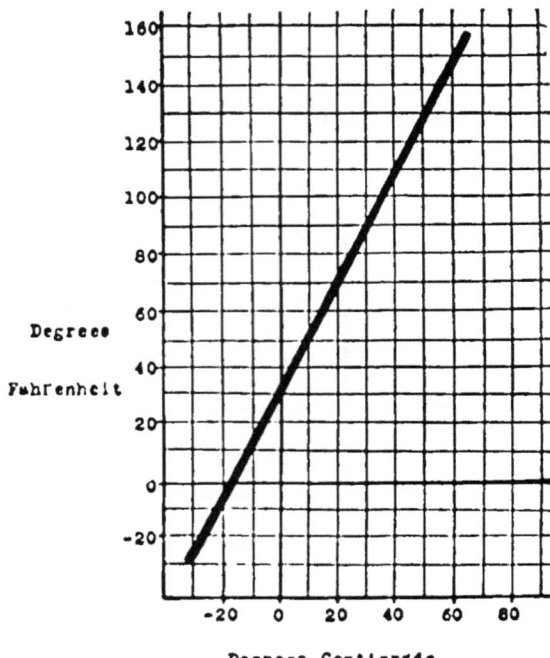

(6) 5. The graph opposite is used to convert degrees of temperature from the Fahrenheit scale (F) to the Centigrade scale (C), and from the Centigrade scale to the Fahrenheit scale.

When it is $+20°$ on the F. scale, what is the temperature in degrees on the C. scale? .....
....

(7) 6. This graph represents the distance passed over by a man walking at the rate of 3 miles an hour for a number of hours.

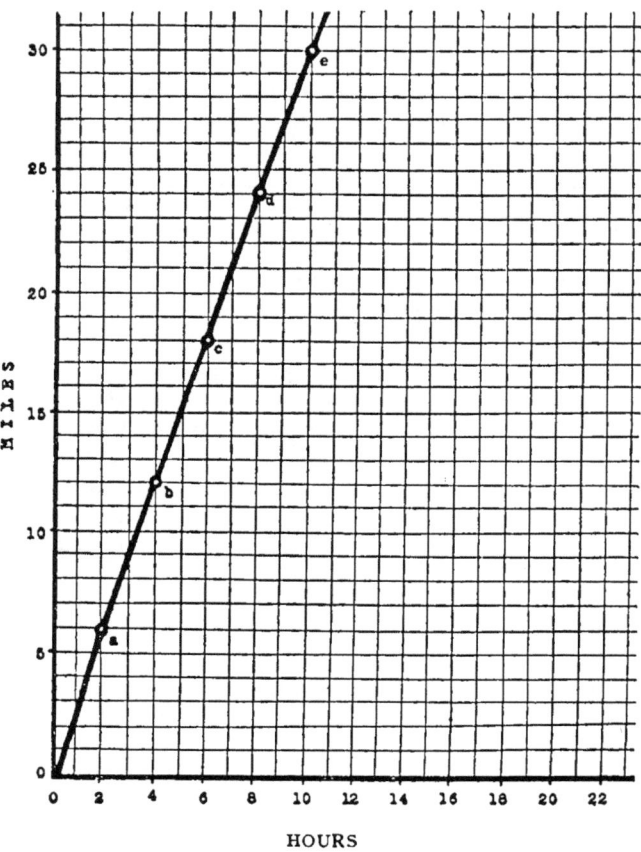

By the conditions of the problem it may be said that the number of miles travelled equals three times the number of hours, or

$$m = 3h$$

That is—
if $h = 2$, then $m = 6$;
if $h = 4$, then $m = 12$;
and if $h = 6$, then $m = 18$, etc.

By locating these points, we have a, b, c, d, etc., of the graph.

In the same diagram draw a graph in which

$$m = 2h$$

(9) 7. Find three pairs of values for *x* and *y* in the following equation and then draw the graph of

$$x + y = 5.$$

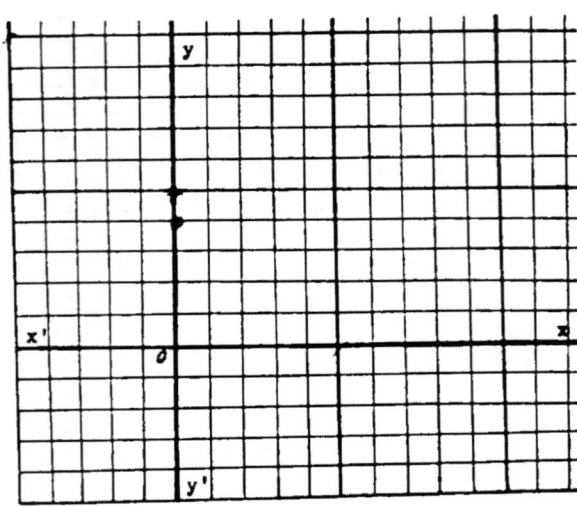

1 Space = 1 Unit

(5) 8. The following table gives the number of inches contained in a number of feet:

Feet	Inches
	12
	24
	36
	48
	60

Represent these facts in the diagram at the right. Let distances on the horizontal line represent feet, 3 spaces equaling 1 foot. Let vertical distances above that line represent inches, 1 space equaling 4 inches. Then locate the points corresponding to the above data and connect by a line.

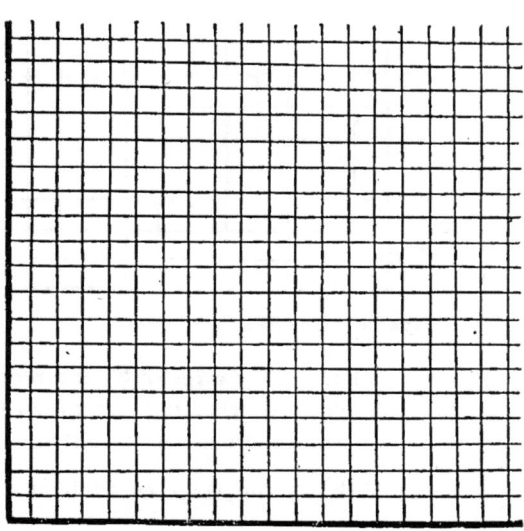

(8) 9. In the diagram for Exercise No. 6, draw the graph of

$$m = 4 + h.$$

10. Plot the following equations and find the values of x and y at the point of intersection of the graphs.

$$x + 4y = 11,$$
$$2x - y = 4.$$

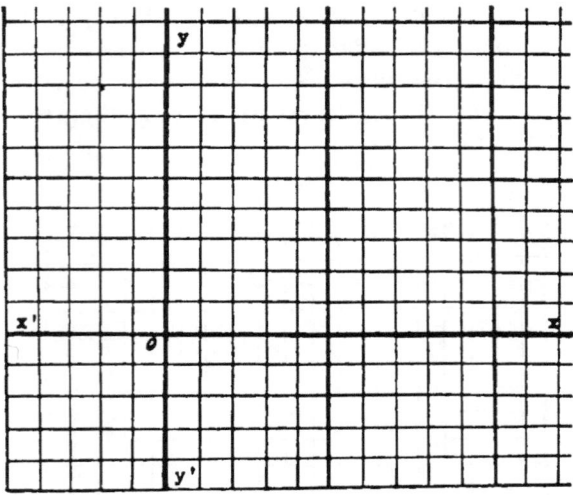

11. A boy begins work with a weekly wage of $9 and receives an increase of 25 cents every week. Another boy starts with a weekly wage of only $6 but receives an increase of 50 cents every week.

Draw a graph which shows the wage of each at the beginning of every week for 15 weeks.

According to this graph when will their wages be the same?.........

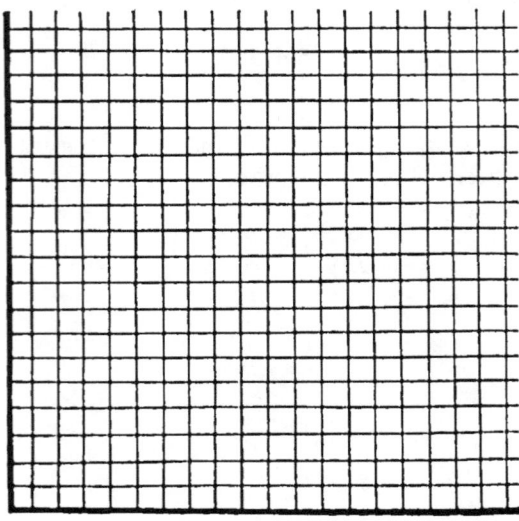

Origin and Use

To insure uniformity to the largest possible extent, all the test sheets were scored by the writer. The exercises were marked either right or wrong according to standards previously determined, and more or less arbitrary. These answers are given in a following section.

II. DIRECTIONS FOR GIVING THE TESTS[3]

Two series of scales are here submitted—Series A and Series B. Series A was derived from Series B. It covers just as wide range of difficulty; and, because it is shorter, it is recommended for use where the time available for testing purposes is limited. Series B includes a richer variety of exercises and problems, thus increasing the diagnostic power of the scales, and is recommended for use wherever time will permit. If only one of the scales can be used, the writer would recommend that the Equation and Formula Scale be used because he feels that it is more comprehensive and tests a much wider function than any of the other scales. If it is desired to use a second scale he would recommend the Problem Scale as coming second in order of importance.

Both series of scales are to be administered in exactly the same way, the only difference being in the amount of time allowed for each test.

In order that results with these scales may be obtained which are fairly comparable, it is absolutely necessary that the same method be pursued in administering the tests as was originally employed in the development of the scales. Directions should be implicitly obeyed. No conversation of any kind should be permitted after a test has once begun, neither by students, teacher, nor person in charge of the test.

While the test sheets are being distributed, ask the students to wait for full directions before doing anything at all with the tests.

When all are ready say to the class, "Now write your name in the first blank space. In the next blank space to your right tell how old you are. Give this in full years to your nearest birthday," etc. After the sheets are properly headed, say to the class, "The exercises on this sheet are in addition and subtraction, collection of terms.[4]

---
[3] See also instructions given in the Appendix, p. 81.
[4] Instead of the phrase 'in addition and subtraction, collection of terms' use 'in multiplication and division', 'on the equation and formula', or 'problems in algebra' whichever the case may be.

Work directly on these sheets. Take the exercises in the order in which they are given. Work as many as you possibly can and be sure you get them right. If you come to one you cannot do, leave it out and pass on to the next."

In the Multiplication and Division Scale call attention to the fact that "all answers must be reduced to their *simplest* forms."

For the Problem Scale add: "In all the problems which call for the equation, *e. g.*, No. 4, simply state the equation which will solve the problem. For example, if you were given this problem— A coat and hat cost $30. The coat cost five times as much as the hat. Find the cost of each. The equation would be $x + 5x = \$30$." (Then write this equation on the board.)

For Graph Scale let the pupils use rulers.

Pupils may be provided with scratch paper for their own use. It has been found to be most satisfactory to pass down the aisles and give each pupil a sheet of scratch paper a few minutes *after* the test has begun. They will find it most convenient, however, to work directly on the question sheets. For all but the Problem Scale it is desirable to have as much of the work as possible on these sheets.

The time limits to be observed in giving the tests are as follows:[5] for each test in Series A allow twenty minutes, except the Problem Scale and the Graph Scale, for each of which twenty-five minutes should be allowed; and for each test in Series B allow forty minutes. By far the greater number of the pupils will have finished before that time, and those who have not will in all probability have solved all they are able to do. A warning, stating the amount of time left, should be given three minutes in advance for the tests of Series A and five minutes in advance for those of Series B.

III. INSTRUCTIONS FOR SCORING AND TABULATING THE RESULTS

As stated before, the exercises or problems were marked either right or wrong. The standards adopted in judging the correctness of an exercise were necessarily quite arbitrary. They were, never-

---

[5] The limits specified above are for classes having studied algebra for a year, or for nine months. When the tests are given to classes having studied algebra for a shorter period of time, the time allowed for each test should be scaled down accordingly.

Origin and Use

theless, determined only after considerable experimentation and personal consultation with teachers of mathematics. These standards are found in Tables I and II. In order to get results which are fairly comparable with those given in this study, it is of course absolutely necessary that the papers be scored in accordance with these standards. To insure greater uniformity in scoring, a few incorrect answers are also included. Reduction to lowest terms was insisted upon only for the test in multiplication and division.

After the papers of a class were scored the results from each class were tabulated upon a separate sheet. A sample of the form[6] as used in this study is given in Fig. 1.

The final value of each exercise or problem for each of the tests is given in Table III, and in the graphic representations which follow this table the difficulty of each of these exercises is projected upon a linear scale.

[6] Inasmuch as the writer was required to repòrt back to each principal the number of exercises or problems solved by each student, together with the median class score, this form was found to be a very convenient source of reference for this purpose as well.

TABLE I
ANSWERS TO EXERCISES

Problem Number	Addition and Subtraction	Multiplication and Division	Equation and Formula
1	$9r$	$21y$	2
2	$5x$	$3n$	3
3	$15b$	$8a^2b^2$	2
4	$2\frac{1}{2}c;\ \dfrac{5c}{2}$	$3c$	7
5	$6x+2$	$6m$	2
6	$8a-6b$	$-2b$	2; not $-z=-2$
7	m	$-12x^2y^3$	9
8	$5x$	$6a^5$	2
9	$r-5t+s$	$2m-3n$	4
10	$5c+6$	$\dfrac{2x^2}{5}$	$\dfrac{15}{16}$
11	a^2-6b+4	$10a^3+33a^2-52a+9$	40
12	$4x-1$	n^2-10	$4\dfrac{2}{7};\ \dfrac{30}{7}$
13	$\dfrac{3c}{8};\ \dfrac{6c}{16}$	$\dfrac{4}{3a}$	15
14	$\dfrac{7x}{6}$	$4x-8;\ 4(x-2)$	$m=2,\ n=4^*$
15	$\dfrac{a-2x}{a^2-x^2}$	$\dfrac{b}{a(m-n)};\ \dfrac{b}{am-an}$	$1\frac{1}{3}$
16	$\dfrac{2r^2}{r^2-z^2};\ \dfrac{2r^2}{(r-z)(r+z)}$	$81x^4y^{12}$	154
17	$\dfrac{7-4a}{6a}$	$c+d;$ not $\dfrac{c^2-d^2}{c-d}$	$\dfrac{Ed}{R}$
18	$\dfrac{9-6x}{4};\ \dfrac{9}{4}-\dfrac{6x}{4}$	$12x^2;\ 12x^{\frac{1}{2}}$	$-\frac{1}{2}$; not $-\frac{1}{2}$; nor $-x=\frac{1}{2}$

TABLE I—(Continued)

Problem Number	Addition and Subtraction	Multiplication and Division	Equation and Formula
19	$\dfrac{3y^2-6x^2}{2x^2\ y^2}$	$a-\tfrac{1}{2}$; $\dfrac{2a-1}{2}$	10, -5 (both roots)
20	$\dfrac{5-3a}{(a+1)(a^2-a+1)}$; $\dfrac{5-3a}{a^3+1}$	$\dfrac{p-2}{3r(p-9)}$; $\dfrac{p-2}{3pr-27}$	$-5, -1$;* not $-x=5$
21	$\dfrac{40-13x}{(x-3)(x-2)(x+5)}$; $\dfrac{40-13x}{x^3-19x+30}$	$\dfrac{x^2-3x+9}{3(x-3)}$; $\dfrac{x^2-3x+9}{3x-9}$	$x=2^*$, $y=4$
22	$\dfrac{1}{(x-1)^3}$; $\dfrac{1}{x^3-3x^2+3x-1}$	48	$21\tfrac{1}{9}$; $\dfrac{190}{9}$
23	$5\tfrac{1}{5}\sqrt{5}$; not $5\sqrt{5}+\tfrac{1}{5}\sqrt{5}$	3	1
24	$\dfrac{6a-7}{a^2-4}$; $\dfrac{7-6a}{4-a^2}$		$\sqrt{\dfrac{2s}{g}}$; $\sqrt{\dfrac{s}{\tfrac{1}{2}g}}$; $\dfrac{1}{g}\sqrt{2gs}$
25			1

* Where two results are ordinarily required in an answer, the exercise was marked correct if the work was done correctly up to the point where only one value was obtained and stopped at that. If, however, the student made an error in solving for the second value, the problem was scored incorrect.

TABLE II
ANSWERS TO PROBLEM TEST

1. $3x$

2. $m - r$

3. $a + b$

4. $x + 10x = 132$. ($12 and $120)

5. $\dfrac{980}{v}$

6. $4x + 40 = 240$; $2x + 2(x-20) = 240$; $4x - 40 = 240$. (50 ft. ×70 ft.)

7. $3x - 136 = 836$; $2x - 136 = 836 - x$; $x + y = 836$ and $x = 2y - 136$.
   (324 children and 512 adults)

8. $\dfrac{r}{dw}$; not $dwx = r$

9. $(x+12)(x-4) = x^2$. (6 ft. ×6 ft. and 2 ft. ×18 ft.)

10. $2\frac{1}{2} : 5\frac{3}{4} = 20 : x$; $x : 20 = 5$ ft. 9 in. : 2 ft. 6 in.; $x : 20 = 69 : 30$; eight times as high as the man; not $2\cdot 6 : 5\cdot 9 = 20 : x$. (46 ft.)

11. $x + y = 5000$ and $\dfrac{3x}{100} + \dfrac{4y}{100} = 172$; $\dfrac{3x}{100} + \dfrac{4(5000-x)}{100} = 172$;
    $.03x + 200 - .04x = 172$. ($2800 and $2200)

12. $40x = 55(x-2)$; $55x = 40(x+2)$; $\dfrac{x}{40} - 2 = \dfrac{x}{55}$. ($293\frac{1}{3}$ miles)

13. $x + y = 20$ and $50x + 65y = 1200$; $50x + 65(20-x) = 1200$; not $50x + 65(20-x) = 12$. ($6\frac{2}{3}$ lbs. and $13\frac{1}{3}$ lbs.)

14. $5x^2 = 180$; $5(x-10)^2 = 180$. (16 in. ×16 in.)

The equations given above are those which are usually found. Modifications which in the end equal the same, may be accepted. For example, $4x = 240 + 40$ is the same as $4x - 40 = 240$, and $\dfrac{69}{30} = \dfrac{x}{20}$ is the same as $69 : 30 = x : 20$. Where the problem has been worked out and the correct answers are given, they are to be scored as correct, though such a procedure on the part of students is to be discouraged.

Origin and Use

TABLE II—(*Continued*)

ANSWERS TO GRAPHS

Problem
No.

        2000

2.   24

3.   3330 to 3350 (inclusive)

4.

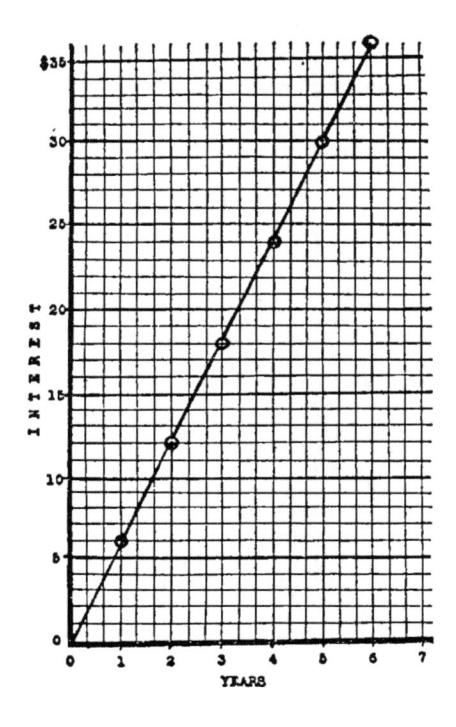

34 *First Year Algebra Scales*

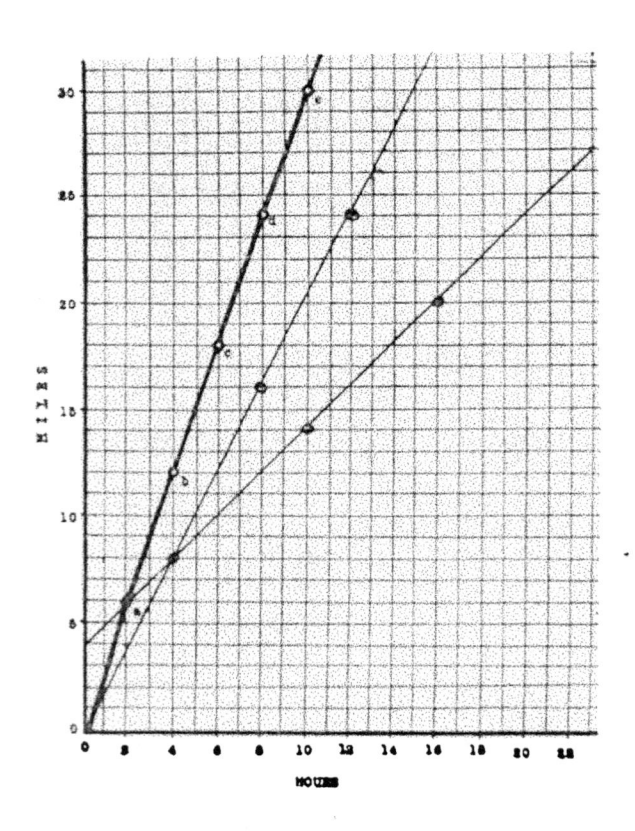

TABLE II—(*Continued*)

ANSWERS TO GRAPHS

Problem
No.

5.   − 3 to − 9 (inclusive)

6.

and

9.

Origin and Use

TABLE II—(Continued)
ANSWERS TO GRAPHS

Problem
No.

7.

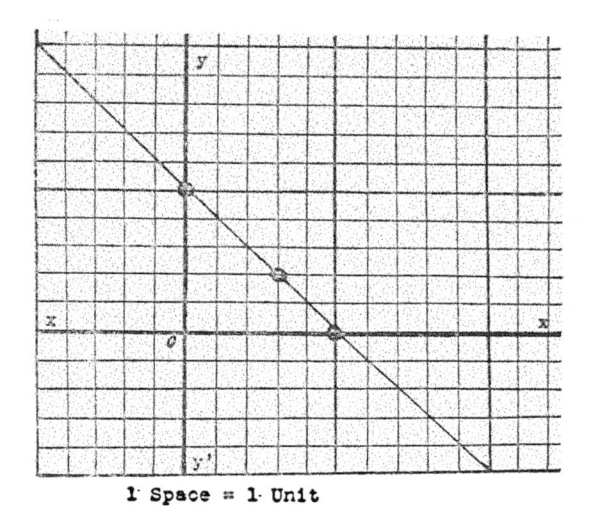

1 Space = 1 Unit

8.

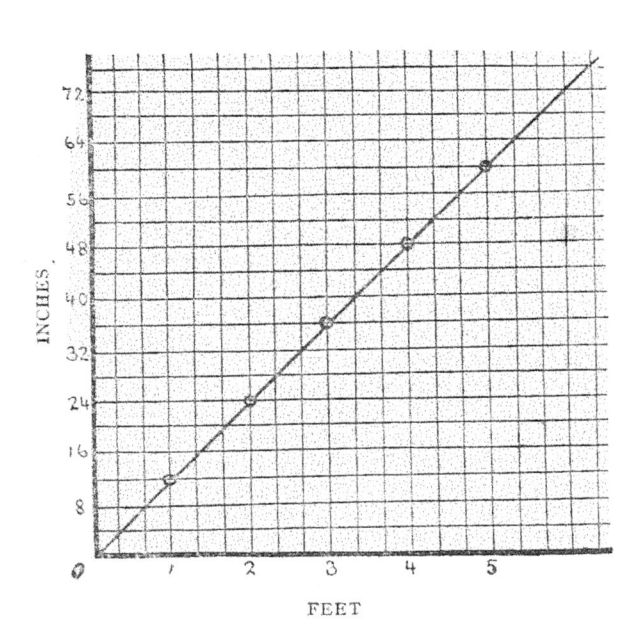

INCHES

FEET

TABLE II—(*Continued*)
ANSWERS TO GRAPHS

Problem No.

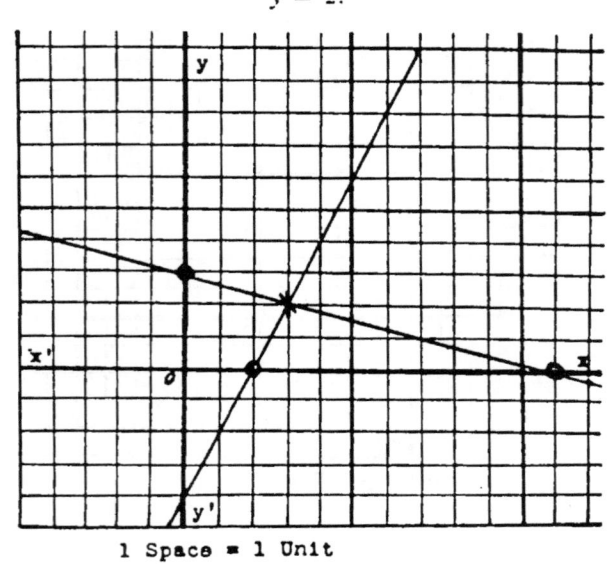

10. $x = 3$
        $y = 2.$

1 Space = 1 Unit

11.   In twelve weeks; the thirteenth week.

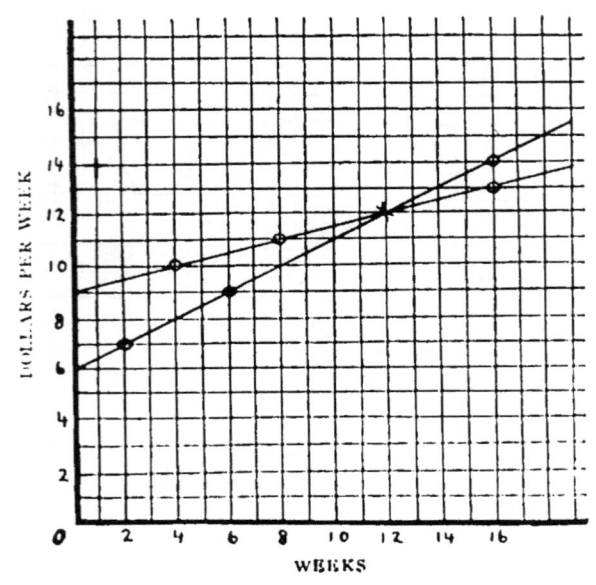

City.............................. School.........................

Test............... Teacher................. Date...............

Problem No.																				No. Correct
1																				
2																				
3																				
4																				
5																				
6																				
7																				
8																				
9																				
10																				
11																				
12																				
13																				
14																				
15																				
16																				
17																				
18																				
19																				
20																				
21																				
22																				
23																				
24																				
25																				
No. Correct																				

Distrib. of Students Acc. to No. of Problems Solved	
0	
1	
2	
3	
4	
5	
6	
7	
8	
9	
10	
11	
12	
13	
14	
15	
16	
17	
18	
19	
20	
21	
22	
23	
24	
25	
No. of Papers	
Class Median	

Figure 1

Sample Score Sheet Used in Tabulating Results from Each Class

TABLE III
Established Value[7] of Each Exercise or Problem in Each Scale

Problem Number	Addition and Subtraction	Multiplication and Division	Equation and Formula	Problems	Graphs
1	*1.17	*1.45	*1.10	*1.07	*1.41
2	1.66	*2.29	*2.27	*2.61	1.98
3	*2.06	*2.92	2.74	3.16	*2.61
4	2.46	3.26	*3.13	*3.35	*3.61
5	*2.97	3.42	3.27	*4.24	*4.46
6	3.19	3.86	*3.49	5.09	*5.06
7	3.30	*3.92	4.13	*5.09	*5.79
8	3.46	4.10	*4.20	5.72	5.97
9	*3.70	*4.54	4.38	*6.20	6.32
10	*4.08	4.93	4.94	6.36	*6.68
11	4.27	*5.14	*4.96	6.42	*7.72
12	*4.55	5.16	5.24	*6.80	
13	*4.92	5.39	5.29	*7.58	
14	5.39	5.44	*5.56	*8.19	
15	*5.61	5.72	5.60		
16	5.74	*5.85	5.99		
17	6.11	*6.60	6.14		
18	*6.41	6.73	*6.17		
19	6.42	6.89	*6.79		
20	6.63	*6.99	6.86		
21	6.70	*7.44	7.15		
22	*6.95	*8.80	7.19		
23	*7.76	*9.32	*7.58		
24	*8.09		*8.53		
25			*9.13		

[7] Values which are starred (*) are for the exercises or problems of Series A.

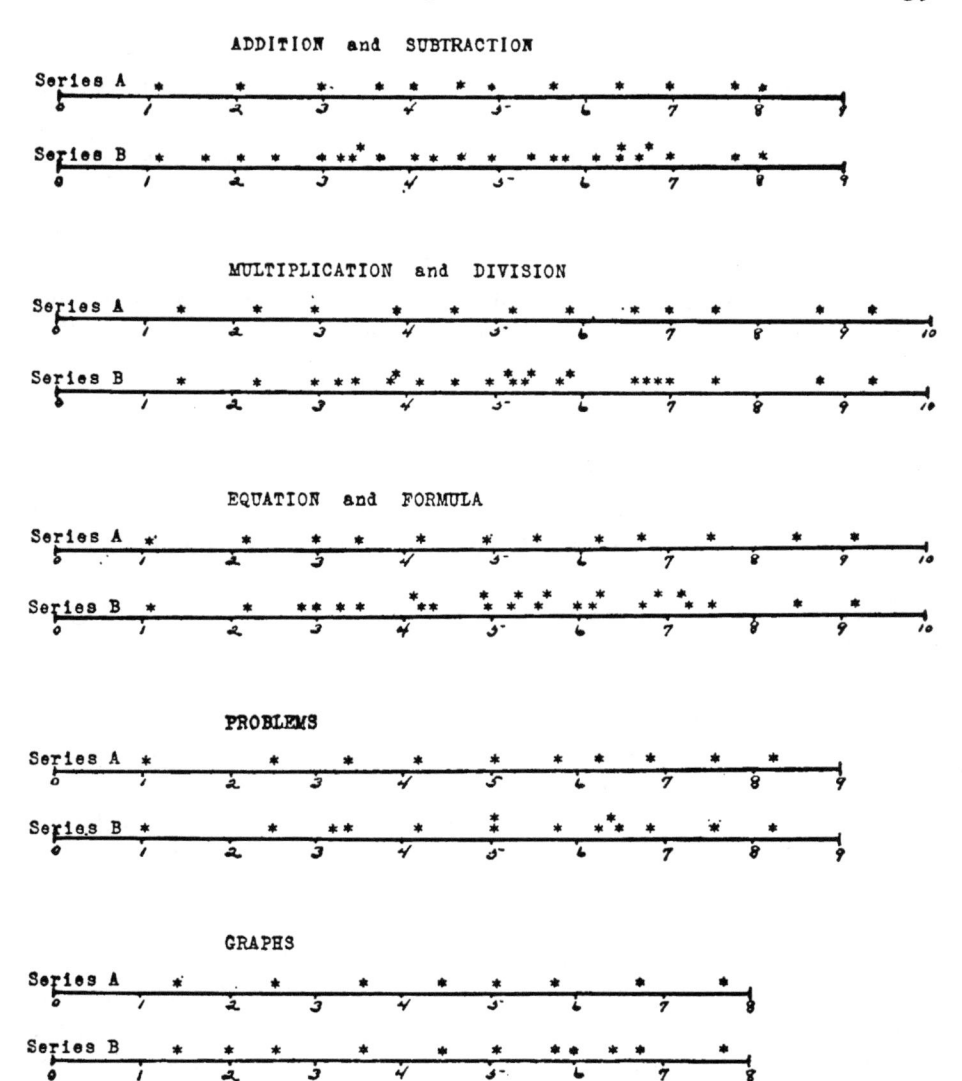

Figure 2
Linear Projection of the Difficulty of the Exercises and Problems,
Series A and Series B

IV. DETERMINATION OF CLASS SCORE

The median class score is recommended for use in connection with these scales as the most satisfactory measurement [8] of the achievement of a class. This median score indicates the point [9] on the scale above and below which there are equal numbers of individual measures, and as such it represents the number of exercises or problems solved correctly by just fifty per cent of the class. That is, there are just as many students in a class who solve a larger number as there are students who solve a smaller number of exercises.

V. TENTATIVE STANDARD SCORES

Tentative standards of achievement based upon the results obtained thus far are given in Tables IV and V. They were calculated upon the basis of the total number of problems solved correctly by each group.

The reliability of such standard scores is determined very largely by the number of pupils tested, and, inasmuch as these numbers are small, it cannot be said that the scores are as yet definitely established. The median scores given for the nine-month group are, however, in all but the Graph Scale derived from the results totaling in the neighborhood of 1,500 cases, as will be seen from tables given farther on. These, it is believed, are quite reliable.

In connection with these scores it should be noted, also, that very few of the classes when tested had studied algebra for the exact time specified, *viz.*, three months, six months, and nine months. Such a procedure, though desirable, would have immeasurably complicated the task of submitting the tests and did not appear feasible. Care was taken, nevertheless, to see that the classes tested tended to group themselves quite evenly about these intervals, and in this way tended to offset the time inequalities which could not be avoided otherwise.

[8] As a measure of the ability of a class when used in connection with the scales of Series B, where the steps between the problems are often very unequal, it is open to rather serious criticism, but because it involves less technical computation it is here adopted as being for all practical purposes the most satisfactory measure.

[9] For instructions as to method of computing median scores see Thorndike, E. L.: *Mental and Social Measurements*, p. 36f; Rugg, H. O.: *Statistical Methods Applied to Education*, p. 100f; Whipple, G. M.: *Manual of Mental and Physical Tests*, p. 13. See also Appendix, p. 81.

TABLE IV
Tentative Median Standards of Achievement, Series A

	Three-Month Group	Six-Month Group	Nine-Month Group
Addition and Subtraction	5.0	6.8	7.9
Multiplication and Division	5.3	6.3	7.9
Equation and Formula	4.9	7.1	7.8
Problem Test	4.3	4.9	5.6
Graph Test	2.8 (four and one-half months)		5.6

TABLE V
Tentative Median Standards of Achievement, Series B

	Three-Month Group	Six-Month Group	Nine-Month Group
Addition and Subtraction	9.7	12.9	14.4
Multiplication and Division	9.6	14.0	16.3
Equation and Formula	7.8	14.3	16.0
Problem Test	5.4	6.5	7.5
Graph Test	3.7 (four and one-half months)		7.2

III

DERIVATION OF THE EQUATION AND FORMULA SCALE

1. ARRANGEMENT OF MATERIAL

The same plan of construction was used in deriving all the scales of this study. It is, therefore, necessary to explain in detail the method pursued in connection with the development of only one of the scales. The Equation and Formula Scale [1] has been selected for this purpose.

The first step taken in the development of this scale was to assemble the results for all the problems of the equation and formula test in two different tables. These crude summaries are given in Tables VI and VII.

Table VI represents the distribution of each of the three groups for the exercises in the Equation and Formula Scale. This table shows that eleven of the students who had studied algebra for three months and one of the students who had studied algebra for six months were unable to solve a single exercise. Nineteen in the first group and one in the second group were able to solve only one exercise, etc. There were 689 students tested who had studied algebra for three months, 746 who had studied algebra for six months, and 1612 who had studied algebra for nine months.

This table also shows the median score for each group, that is, the number of exercises such that there are just as many students in the group who solve a greater number of exercises as there are students who solve a less number. The difference between the twenty-five percentile and the seventy-five percentile shows us the range in the number of exercises solved by the middle fifty per cent of the students of a group. One half of this difference gives us the

---

[1] All the essential data from which the other scales were derived are given in more condensed form in the last section. From these tables, after following the explanations given here, it will be comparatively easy to see how the final value of each problem was computed.

TABLE VI
Distribution According to the Number of Equation and Formula Exercises Solved

Number Solved	Three-Month Group	Six-Month Group	Nine-Month Group
0	11		
1	19		
2	24	3	4
3	26	3	1
4	37	6	6
5	63	8	10
6	87	19	14
7	98	31	18
8	93	37	30
9	102	36	58
10	44	48	72
11	37	53	106
12	19	57	106
13	11	47	127
14	9	77	139
15	4	62	115
16	3	77	148
17	1	60	135
18		56	134
19		28	125
20		16	93
21		10	78
22		7	44
23		2	23
24		1	12
25			4
Number Tested	689	746	1612
Median [2]	7.791	14.299	16.000
25 Percentile	5.877	10.865	12.792
75 Percentile	9.576	16.915	18.895
Quartile	1.849	3.025	3.051

[2] In ordinary practice it is not necessary to work out the medians and percentiles farther than the first decimal place. To carry these out to three decimal places, as is often done in this study, suggests a degree of accuracy and refinement in statistics which these scales do not possess.

quartile, or semi-interquartile range, which is a measure of the variability of the group.

Table VII indicates the number of students from each group that solved each exercise correctly. Thus, 665 of the students who had studied algebra for three months, 741 of the students who had studied algebra for six months, and 1,608 of those who had studied algebra for nine months, succeeded in getting the first exercise, etc.

TABLE VII

Number in Each Group That Solved Each Exercise
in Equation and Formula Test Correctly

Problem Number	Three-Month Group	Six-Month Group	Nine-Month Group
	665	741	1608
	564	714	1552
3	597	728	1580
4	516	693	1504
5	465	662	1452
6	449	687	1496
7	328	581	1285
8	373	567	1284
9	275	356	1110
10	333	518	1264
11	72	538	1290
12	119	427	1031
13	43	510	1133
14	17	410	1106
15	90	209	776
16	5	427	1121
17	5	299	830
18	9	274	863
19	5	155	572
20	1	144	528
21		65	286
22		122	397
23		101	408
24		8	123
25		4	46
No. Tested	689	746	1612

II. PROBABLE ERROR (P. E.) TAKEN AS THE UNIT OF MEASURE

Scales aim to establish units of measurement which shall be both accurate and constant. In them values are scientifically established, standard conditions are maintained, and the personal element, which has played such a prominent part in the educational measurements in the past, is reduced to a minimum. Thus, we are able to get results which can be universally interpreted and compared. As instruments of measurement, scales are, therefore, infinitely superior to the snap judgments of teachers, even after such judgments have been corrected by the results of a series of tests wherein the exact difficulty of each problem is necessarily more or less of an unknown quantity.

In the construction of these algebra scales an attempt has been made to develop reliable objective instruments for measuring the ability of first year algebra students, instruments that can be used by different people in making measurements and yet get results that are comparable. The difficulty of each exercise or problem has been definitely established and its position above a selected zero point determined. That is, the exercises have all been placed in their relative positions on a projected linear scale. In Series A, special effort was made to select exercises with equal steps of difficulty between them.

Points and steps upon scales for measuring school products may be determined in one of two ways. They may be based upon the judgments of a number of persons competent to judge the order of difficulty of a series of problems; or, as was done in this study, such exercises or problems may be submitted to a number of pupils and then scaled in order of difficulty as determined by these results.

The unit used in constructing these algebra scales is what is commonly known in statistical measurements as the median deviation or probable error (P. E.). It is a measure of variability and is the unit generally selected in the construction of scales because it is the most constant measure known.[3]

Before going further into the discussion of the significance of P. E. as our unit of measure, it is, however, necessary to state here the fundamental assumptions upon which such a procedure depends. It assumes first, that the ability of pupils to solve the

---

[3] For a more complete discussion of this question, consult Trabue, Marion Rex: *Completion-Test Language Scales*, p. 30.

algebra exercises or problems is distributed within the respective groups (three-month group, six-month group, and nine-month group) according to the normal surface of distribution; and in the second place, it assumes that the variability of any one group is equal to that of any other group.

Neither of these assumptions may be strictly valid, nor have they been scientifically established by the evidence which is available. It is, however, believed that for all practical purposes they offer the most reliable working basis which could be adopted at present.[4]

According to the first of these assumptions—that the achievement of any group in the solution of algebra exercises or problems tends

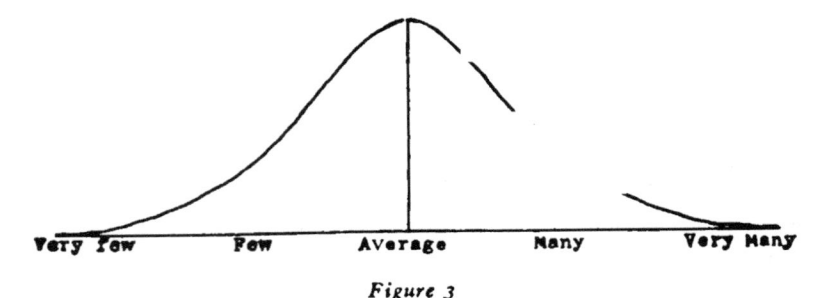

Very Few   Few   Average   Many   Very Many

Figure 3
Normal Surface of Frequency Showing the Distribution of Achievement in the Solution of Exercises or Problems.

to distribute itself normally—we should expect to find that most of the students solve about the same number of exercises, and that they cluster about the median or average for that group. Furthermore, we should expect that there is a relatively small number of students who can solve only a few exercises and, likewise, only a few who can solve many more than the average number of exercises.

[4] A few psychologists maintain that by the time the first year of the high school is reached the forces of selection have already operated to such an extent as to affect quite decidedly the normal distribution of ability to do scholastic work. They believe that elimination has taken place very largely in the lower half of the curve, and that our distribution should, therefore, be skewed just a little to the left.

On the other hand, there are others who deny that such a positive skew really exists. They argue that by the time pupils reach the high school elimination has been due to a variety of interrelated causes, a number of which have nothing whatever to do with intellectual ability.

That some selection on the basis of scholarship does take place may well be admitted, but on the whole it is as yet questionable whether this selection up to and into the high school is of such a nature as to change the form of the distribution to a marked degree.

The Equation and Formula Scale

In general, we should expect to find the achievement of these students distributed symmetrically on each side of the average.

Fig. 3 illustrates such a distribution. All the students arranged according to the number of exercises solved are represented by the space enclosed between the curve and the base line. The height of the curve above the base line indicates the number of individuals of the group that have the ability shown on the base line

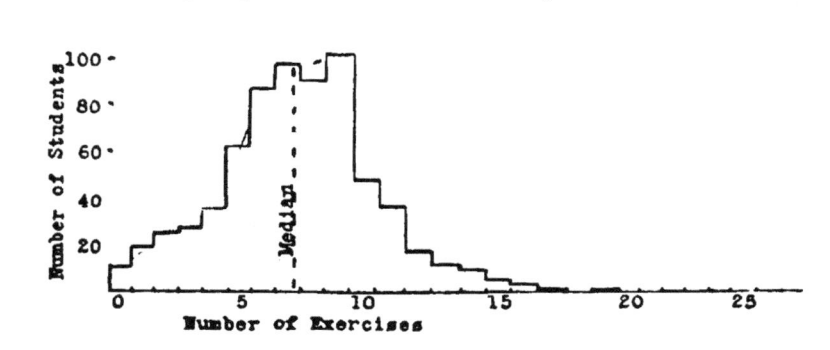

Figure 4
Distribution According to Number of Exercises Solved by Three-Month Group in Equation and Formula Test.

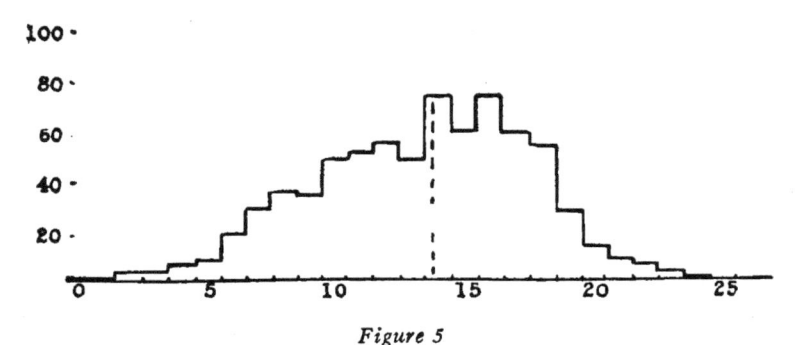

Figure 5
Distribution According to Number of Exercises Solved by Six-Month Group in Equation and Formula Test.

scale, and each individual is represented by an equal amount of the enclosed area. Thus, at the extreme left, the curve is very near the base, which shows that the number of students who could solve only a few exercises was small. In the middle the curve is at a maximum distance from the base line, indicating the large number of students who solved an average number of problems, while at the extreme right the curve is again very near the base, representing the small number of students who were able to solve a large number of problems.

With these characteristics of the normal distribution curve in mind, it is interesting to note to what extent the achievement of students as shown by the exercises in the Equation and Formula test approximated the normal curve of distribution. Figs. 4, 5, and 6 represent graphically the distribution of the various groups tested by the equation and formula exercises.

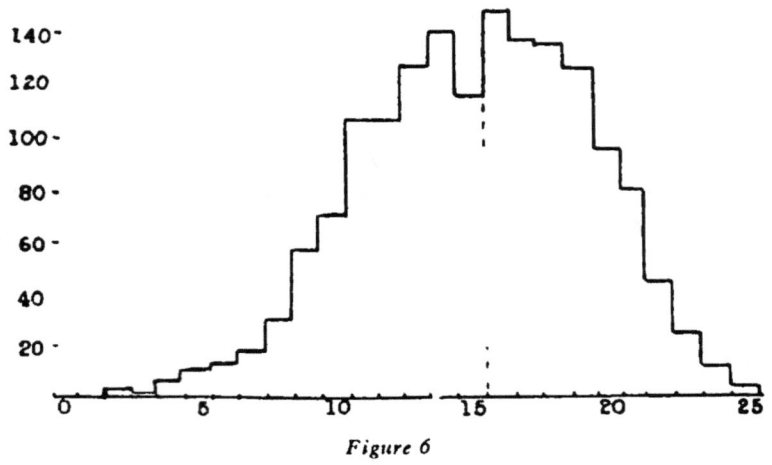

Figure 6

Distribution According to Number of Exercises Solved by Nine-Month Group in Equation and Formula Test.

It will be seen from these figures that the three-month distribution is skewed positively just a little; and the nine-month distribution, on the other hand, has a slight negative skew. In general, however, the figures bear a closer resemblance to the normal frequency curve than we would be led to expect.[5]

We are now prepared to define the unit of variability which has been adopted as the most convenient unit for the measurement

---

[5] Several factors which have a tendency to distort the form of the distribution were operating. In the first place, the exercises used did not prove, as has already been seen (Fig. 2), to be such that they could be arranged upon a linear scale differing in difficulty by approximately equal amounts. That is, the exercises did not increase in difficulty by equal intervals, nor was there a sufficient number of very easy and of very difficult exercises in the test. In the second place, we cannot be certain that the groups were strictly speaking non-select with regard to intellectual ability.

The Equation and Formula Scale 49

of achievement in the construction of these scales. This, as stated above, is the median deviation or probable error and is usually designated by the letters P. E. It is defined as the median amount of deviation from the central tendency, and can perhaps be more satisfactorily explained in connection with Fig. 7. In this figure the perpendicular *mn* divides the normal surface of distribution into two equal parts, so that fifty per cent of the cases are on each side of this perpendicular. The point where the perpendicular cuts the base line is the median point. To the left of this point the perpendicular *ab* is drawn so as to include twenty-five per cent of the whole number of cases between this perpendicular and the median perpendicular.

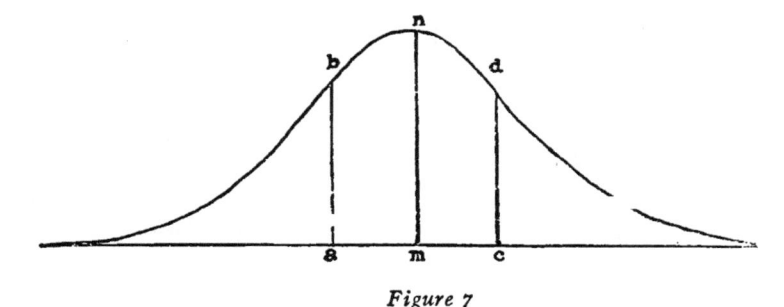

Figure 7
Normal Surface of Distribution Showing the Median and the P. E. Distance on Each Side of the Median Point.

Likewise, the perpendicular *cd* includes twenty-five per cent of the cases to the right of the median perpendicular. It will be seen that the area *abcd* contains the middle fifty per cent of all the cases. The distance *ma* or *mc* on the base line of the surface of distribution is the median deviation or P. E.

Furthermore, it has been established that if a perpendicular be erected at a distance of 2 P. E. on the base line of the surface of frequency, and on either side of the median point, it will include with the median perpendicular 41.13 per cent of all the cases; if erected at a distance of 3 P. E. it will include 47.85 per cent of the cases; and if erected at a distance of 4 P. E. it will include 49.65 per cent of the cases. In theory the curve and base line never meet and could be indefinitely extended, always getting nearer together but never touching. For the purposes of this study it has been considered sufficiently accurate to assume that they meet at a distance of 4.6 P. E. from the median, since the perpendicular erected here on either side of the median cuts off only 0.1 per cent of all the cases.

III. SCALING THE EXERCISES IN THE EQUATION AND FORMULA TEST FOR EACH GROUP

Having adopted the P. E. of the distribution of a group as our unit of measure and having assumed that the ability of students in solving the equation and formula exercises is distributed according to the normal distribution curve, our next task is to locate each

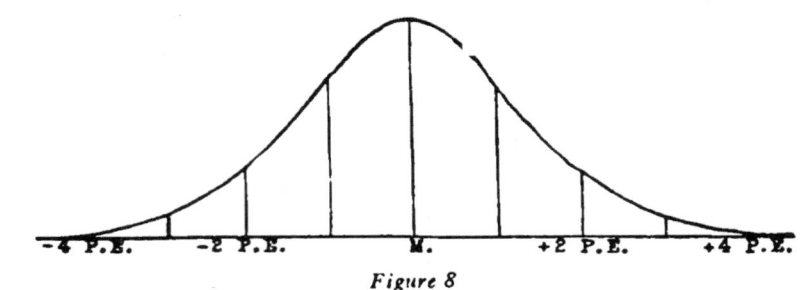

Figure 8
Normal Surface of Frequency Showing P. E. Distances from the Median Point.

exercise on the base line of each group distribution. An exercise that is solved correctly by just fifty per cent of the students of a group would represent the median achievement of the group and would very obviously be located at the median point on the base line. Since 1 P. E. represents the distance along the base line, on either side of the median, to a point where a perpendicular erected will include between this perpendicular and the median twenty-five per cent of the cases, it is evident that an exercise which is solved by only twenty-five per cent of the students is just one unit more difficult than the one just mentioned and should be located at +1 P. E. from the median point. Similarly, an exercise that is solved by seventy-five per cent of the group is considered as being one unit less difficult and so it is located at —1 P. E. from the median. In this way after the per cent of students of a group that solved a certain exercise is known, the deviation of such exercise in per cents from the median or fifty per cent can be calculated. Such values may then in turn be translated into P. E. distances from the median achievement of the group, and thus a P. E. scale for each group can be constructed.

TABLE VIII

Per Cent of Each Group That Solved Each Exercise in Equation and Formula Test

Problem Number	Three-Month Group	Six-Month Group	Nine-Month Group
	96.52	99.33	99.75
	81.9	95.7	96.3
3	86.7	97.6	98.0
4	74.9	92.9	93.3
5	67.5	88.7	90.1
6	65.2	92.1	92.8
7	47.6	77.9	79.7
8	54.1	76.0	79.6
9	39.9	47.7	68.9
10	48.3	69.4	78.4
11	10.5	72.1	80.0
12	17.3	57.2	63.8
13	6.2	68.4	70.3
14	2.5	55.0	68.6
15	13.1	28.0	48.1
16	.7	57.2	69.5
17	.7	40.1	51.5
18	1.3	36.7	53.5
19	.7	20.8	35.5
20	.2	19.3	32.7
21		8.7	17.7
22	.4	16.3	24.6
23	.3	13.5	25.3
24		1.1	7.6
25		.53	3.47

Table VIII has been constructed from Table VII (page 44). It gives the percentage of each group that solved each of the equation and formula exercises. For example, from Table VII we note that 665 out of a total of 689 students in the three-month group solved the first exercise. Table VII gives this as 96.52[6] per cent.

[6] It is ordinarily necessary to carry this out to only one decimal place but inasmuch as the corresponding P. E. distances upon the base line become increasingly large at the extreme ends of the distribution such values are here calculated to two decimal places.

TABLE IX

Difference Between Fifty Per Cent and the Per Cent in Each Group That Solved Each Exercise in Equation and Formula Test

Problem Number	Three-Month Group	Six-Month Group	Nine-Month Group
	46.52	49.33	49.75
	31.9	45.7	46.3
3	36.7	47.6	48.0
4	24.9	42.9	43.3
5	17.5	38.7	40.1
6	15.2	42.1	42.8
7	−2.4	27.9	29.7
8	4.1	26.0	29.6
9	−10.1	−2.3	18.9
10	−1.7	19.4	28.4
11	−39.5	22.1	30.0
12	−32.7	7.2	13.8
13	−43.8	18.4	20.3
14	−47.5	5.0	18.6
15	−36.9	−22.0	−1.9
16	−49.3	7.2	19.5
17	−49.3	−9.9	1.5
18	−48.7	−13.3	3.5
19	−49.3	−29.2	−14.5
20	−49.8	−30.7	−17.3
21		−41.3	−32.3
22	−49.6	−33.7	−25.4
23	−49.7	−36.5	−24.7
24		−48.9	−42.4
25		−49.47	−47.15

Table IX shows the deviation from the median achievement of each group. It is obtained by subtracting fifty per cent from the per cents given in Table VIII.

Table X has been calculated mathematically [7] and is introduced here to be used in connection with Table IX. It indicates the corresponding P. E. value for each 0.1 per cent deviation from the median.

[7] Table X is taken directly from *Spelling Ability*, Table XLVIII, by B. R. Buckingham. It is a revision of the table given in *Mental and Social Measurements*, p. 198, by E. L. Thorndike.

TABLE X
P. E. Values Corresponding to Given Per Cents of the Normal Surface of Frequency, Per Cents Being Taken From the Median

	0	.1	.2	.3	.4	.5	.6	.7	.8	.9
0	.000	.004	.007	.011	.015	.019	.022	.026	.030	.033
1	.037	.041	.044	.048	.052	.056	.059	.063	.067	..071
2	.074	.078	.082	.085	.089	.093	.097	.100	.104	.108
3	.112	.115	.119	.123	.127	.130	.134	.138	.141	.145
4	.149	.153	.156	.160	.164	.168	.172	.175	.179	.183
5	.187	.190	.194	.198	.201	.205	.209	.213	.216	.220
6	.224	.228	.231	.235	.239	.243	.246	.250	.254	.258
7	.261	.265	.269	.273	.277	.280	.284	.288	.292	.296
8	.299	.303	.307	.311	.315	.318	.322	.326	.330	.334
9	.337	.341	.345	.349	.353	.357	.360	.364	.368	.372
10	.376	.380	.383	.387	.391	.395	.399	.403	.407	.410
11	.414	.418	.422	.426	.430	.434	.437	.441	.445	.449
12	.453	.457	.461	.464	.468	.472	.476	.480	.484	.489
13	.492	.496	.500	.504	.508	.512	.516	.519	.523	.527
14	.531	.535	.539	.543	.547	.551	.555	.559	.563	.567
15	.571	.575	.579	.583	.588	.592	.596	.600	.603	.608
16	.612	.616	.620	.624	.628	.632	.636	.640	.644	.648
17	.652	.656	.660	.665	.669	.673	.677	.681	.685	.689
18	.693	.698	.702	.706	.710	.714	.719	.723	.727	.731
19	.735	.740	.744	.748	.752	.756	.761	.765	.769	.773
20	.778	.782	.786	.790	.795	.799	.803	.807	.812	.816
21	.820	.825	.829	.834	.838	.842	.847	.851	.855	.860
22	.864	.869	.873	.878	.882	.886	.891	.895	.900	.904
23	.909	.913	.918	.922	.927	.931	.936	.940	.945	.949
24	.954	.958	.963	.968	.972	.977	.982	.986	.991	.996
25	1.000	1.005	1.009	1.014	1.019	1.024	1.028	1.033	1.038	1.042
26	1.047	1.052	1.057	1.062	1.067	1.071	1.076	1.081	1.086	1.091
27	1.096	1.101	1.105	1.110	1.115	1.120	1.125	1.130	1.135	1.140
28	1.145	1.150	1.155	1.160	1.165	1.170	1.176	1.181	1.186	1.191
29	1.196	1.201	1.206	1.211	1.217	1.222	1.227	1.232	1.238	1.243
30	1.248	1.253	1.259	1.264	1.269	1.275	1.279	1.286	1.291	1.296
31	1.302	1.307	1.313	1.318	1.324	1.329	1.335	1.340	1.346	1.351
32	1.357	1.363	1.368	1.374	1.380	1.386	1.391	1.397	1.403	1.409

TABLE X—(Continued)
P. E. Values Corresponding to Given Per Cents of the Normal Surface of Frequency, Per Cents Being Taken from the Median

	0	.1	.2	.3	.4	.5	.6	.7	.8	.9
33	1.415	1.421	1.427	1.432	1.438	1.444	1.450	1.456	1.462	1.469
34	1.475	1.481	1.487	1.493	1.499	1.506	1.512	1.518	1.524	1.531
35	1.537	1.543	1.549	1.556	1.563	1.569	1.576	1.582	1.589	1.595
36	1.602	1.609	1.616	1.622	1.629	1.636	1.643	1.649	1.656	1.663
37	1.670	1.677	1.685	1.692	1.699	1.706	1.713	1.720	1.728	1.735
38	1.742	1.749	1.757	1.765	1.772	1.780	1.788	1.795	1.803	1.811
39	1.819	1.827	1.835	1.843	1.851	1.859	1.867	1.875	1.884	1.892
40	1.900	1.909	1.918	1.926	1.935	1.944	1.953	1.962	1.971	1.979
41	1.988	1.997	2.007	2.016	2.026	2.035	2.044	2.054	2.064	2.074
42	2.083	2.093	2.103	2.114	2.124	2.134	2.145	2.155	2.166	2.177
43	2.188	2.199	2.211	2.222	2.234	2.245	2.257	2.269	2.281	2.293
44	2.305	2.318	2.331	2.344	2.357	2.370	2.384	2.397	2.411	2.425
45	2.439	2.453	2.468	2.483	2.498	2.514	2.530	2.546	2.562	2.579
46	2.597	2.614	2.631	2.648	2.667	2.686	2.706	2.726	2.746	2.767
47	2.789	2.811	2.834	2.857	2.881	2.905	2.932	2.958	2.986	3.015
48	3.044	3.077	3.111	3.146	3.182	3.219	3.258	3.300	3.346	3.395
49	3.450	3.506	3.571	3.643	3.725	3.820	3.938	4.083	4.275	4.600
50										

Table XI gives the P. E. value of each exercise for each group distribution, and so indicates the position on the base line with reference to the median achievement of each exercise for each distribution. From this table, then, the difficulty of any exercise for any group may be found.

To assist in a fuller interpretation of these tables as used in finding the difficulty of an exercise by locating it on the base line of one of the group distributions, let us find the position of the second exercise as determined by the achievement of the nine-month group. By referring to Table VII, we find that 1,552 of the 1,612 students who tried exercise No. 2 solved it correctly. Table VIII shows that this is 96.3 per cent of the total number who tried it, or, according to Table IX, 46.3 per cent more than the median achieve-

TABLE XI

P. E. Equivalent of Difference Between Fifty Per Cent and the Per Cent in Each Group That Solved Each Exercise in Equation and Formula Test

Problem Number	Three-Month Group	Six-Month Group	Nine-Month Group
	−2.688	−3.668	−4.159
	−1.351	−2.546	−2.648
3	−1.649	−2.932	−3.044
4	−0.996	−2.177	−2.222
5	−0.673	−1.795	−1.909
6	−0.579	−2.093	−2.166
7	0.089	−1.140	−1.232
8	−0.153	−1.047	−1.227
9	0.380	0.085	−0.731
10	0.063	−0.752	−1.165
11	1.859	−0.869	−1.248
12	1.397	−0.269	−0.523
13	2.281	−0.710	−0.790
14	2.905	−0.187	−0.719
15	1.663	0.864	0.071
16	3.643	−0.269	−0.756
17	3.643	0.372	−0.056
18	3.300	0.504	−0.130
19	3.643	1.206	0.551
20	4.275	1.286	0.665
21		2.016	1.374
22	3.938	1.456	1.019
23	4.083	1.636	0.986
24		3.395	2.124
25		3.791	2.822

ment. The corresponding P. E. value of 46.3 per cent as obtained from Table X is 2.648; and since this exercise is just that much too easy when compared with the median achievement of the nine-month group, it is located this distance below the median, or as seen in Table XI at −2.648 P. E. The same exercise is located, as can be seen from the same table, at −2.546 P. E. in the six-month distribution and at −1.351 P. E. in the three-month distribution.

IV. MEASUREMENT OF DISTANCES BETWEEN THE MEDIANS OF SUCCESSIVE GROUPS

We have now succeeded in locating each exercise at its proper distance from the median point of each group distribution. That is, we have established a P. E. scale for each of the three groups. It is our purpose ultimately, however, to locate each exercise upon one linear scale so as to know how difficult each exercise is in general. To do so we must next find the P. E. distance between the consecutive groups.

Three methods were used in this study to determine these intergroup intervals. They are here designated as the 'problem method', the 'quartile method', and the 'overlapping method'.

By the problem method the distance between two consecutive groups is determined by the difference in position which each exercise holds within each respective group with reference to the median. It will be seen that this method not only assumes that the achievement of students in solving the exercises is distributed according to the normal frequency curve, but, as may also be inferred from the above, that their achievement in solving each individual exercise tends to be likewise distributed. The intervals are calculated from Table XI. According to this table, the third exercise has a P. E. value for the three-month group of -1.649, and in the six-month group it has a P. E. value of -2.932. The difference between these two values gives us 1.283, which is the P. E. distance between these two groups as determined by this exercise. Each exercise gives us a separate measure of the interval between successive groups. These intervals have been calculated for all the exercises in the equation and formula test and are recorded in Table XII.

In connection with this table, some attention must be given to the fact that the more difficult the exercise the larger the interval between two consecutive groups tends to become. That this is true is more clearly shown for these exercises in Table XIII. In this table the intergroup intervals as determined by the various exercises are classified roughly according to the difficulty of the exercise. The results marked below -1.5 P. E. are the averages of those determinations in Table XII which came from values any one or both of which in Table XI were lower than -1.5 P. E.;

the results marked −1.5 P. E. to +1.5 P. E. are the averages of those determinations which came from values both of which were between −1.5 P. E. and +1.5 P. E.; and the results marked above +1.5 P. E. are the averages of those determinations which came from values any one or both of which were above +1.5 P. E. The results marked 'select class' are the averages of those determinations only which were derived from values between −2 P. E. and +2 P. E., while the 'composite average' is the average of all the determinations given in Table XII.

TABLE XII
P. E. Interval Shown Between Consecutive Groups
by Each Equation and Formula Exercise

Problem Number	Three- to Six-Month Interval	Six- to Nine-Month Interval
	.980	.491
	1.195	.102
3	1.283	.112
4	1.181	.045
5	1.122	.114
6	1.514	.073
7	1.229	.092
8	.894	.180
9	.295	.816
10	.815	.413
11	2.728	.379
12	1.666	.254
13	2.991	.080
14	3.092	.532
15	.799	.793
16	3.912	.487
17	3.271	.428
18	2.796	.634
19	2.437	.655
20	2.989	.621
21		.642
22	2.482	.437
23	2.447	.650
24		1.271
25		.969

TABLE XIII

Average of Various Determinations of Intergroup Intervals as Measured by the Equation and Formula Exercises

	Three- to Six-Month Interval	Six- to Nine-Month Interval
P. E. Value Below −1.5	1.214	.156
−1.5 P. E. to +1.5 P. E.	.980	.453
Above +1.5 P. E.	2.722	.883
Select Class	1.194	.445
Composite Average	1.914	.451

The results as a whole show that each group is superior to the one next below it in the solution of all exercises, but the largest differences are brought about by the difficult exercises.[3] One exception to this general rule occurs in Table XIII, and it is the only one found in this entire study. It will be seen that the average interval of the exercises −1.5 P. E. to +1.5 P. E. happens to be smaller for the three- to six-month interval than it is for the easier exercises, that is, for those below −1.5 P. E. This may be explained in the first place by the fact that only five cases fell between these limits, and in the second place three of these exercises happened to be almost as easy for the three-month group as they were for the nine-month group.

Several plans could be pursued in making the final determinations of the intergroup intervals by the 'problem method'. The easy exercises tend to give us a small intergroup interval, while the difficult ones tend to increase this difference between the medians of two consecutive groups. Of these two types of exercises the difficult ones probably give us the more reliable measure; but, on the whole, it would seem that the best measure of the interval is the average of those determinations that came from exercises near the median. These should, therefore, be weighted while those exercises which are more than 3 or 4 P. E. from the median should perhaps not be allowed to influence the measure of the interval at all. In this study both the composite average and the average of those de-

[3] This fact was noted by Dr. Trabue and also by Dr. Woody in their scale studies

The Equation and Formula Scale 59

terminations given in Table XII which were obtained from values −2 P. E. to +2 P. E. in Table XI were used in making the final calculations. The latter is designated in Table XIII as the 'select class'; and, inasmuch as this measure is in all probability more reliable than the composite average, it has seemed advisable to give it double weight. The final determinations of the intergroup intervals by the 'problem method' are therefore as follows:

	Three- to Six- Month Interval	*Six- to Nine- Month Interval*
Select Class	1.194	.445
Select Class	1.194	.445
Composite Average	1.914	.451
Average	*1.434*	*.447*

The second method adopted to get a measure of the distance between the group medians has been designated as the 'quartile method'. This method uses the quartile as a measure of P. E. The quartile as we have already noted in connection with Table VI is one-half of the range of the middle fifty per cent. If then, as we have assumed in this study, we have a normal surface of distribution, the quartile is exactly equal to the P. E. We can, therefore, divide our crude score interval by the quartile of the distribution and get another measure of our intergroup intervals in terms of P. E.

Table XIV shows the intergroup intervals obtained by this process.

This table is constructed from data taken from Table VI. The crude score interval is the difference between the median scores. That is, the difference between the median number of problems solved by successive groups. Since there are two quartile measures for each crude score interval, the average of the two is taken as the divisor of the crude score interval. Thus, to get in terms of P. E. the measure of the distance between the medians of the three-month and the six-month groups by the 'quartile method', we subtract 7.791 from 14.299. This gives us the crude score interval of 6.508.

The average quartile is then found by taking the average of the quartiles 1.849 and 3.025, which is 2.437. Finally, the quartile interval is calculated by dividing 6.508 by 2.437, which is 2.671.

TABLE XIV
Determination of Quartile Intervals Between the Groups

	Three-Month Group	Six-Month Group	Nine-Month Group
Median	7.791	14.299	16.000
Quartile	1.849	3.025	3.051
Crude Score Interval		6.508	1.801
Average Quartile		2.437	3.038
Quartile Interval		2.671	.592

The third method of calculating the intergroup intervals is based upon the amount of overlapping by consecutive group distributions. By this method that part of the group distribution which lies between its own median and that of another is translated directly into a P. E. value, and so gives us another measure of the distance between group medians.

Reference to Table VI reveals the fact that many students of the three-month group surpassed the median score of the six-month group; and, conversely, that many students of the six-month group fell below the median score of the three-month group. Between the median of the first group, which is 7.791, and that of the second group, which is 14.299, there are (assuming that the scores within each step are spread evenly over that step) 329.13 of the 689 scores, or 47.77 per cent of the individuals in the three-month distribution. Since this per cent is a deviation from the median or fifty per cent, it can be turned directly into a P. E. value by the aid of Table X. Its P. E. value is 2.978. This is known in Table XVI as the 'lower direct' measure of the three- to six-month interval. Similarly, it is found that 41.22 per cent of the individuals of the six-month distribution are included between the medians 7.791 and 14.299, and the P. E. equivalent of 41.22 per cent is 2.009.

This is known as the 'upper direct' measure of the three- to six-month interval.

It is possible also to obtain an indirect measure of this same interval. Between the three-month median and the nine-month median lies 46.95 per cent of the nine-month distribution. Thus, referring again to Table X, we may say that the three-month median, as measured by the nine-month distribution, is 2.780 P. E. below the nine-month median. Likewise, between the nine-month median and the six-month median, there is included 13.18 per cent of the nine-month distribution, which makes the P. E. distance

TABLE XV

Percentages, with P. E. Equivalents, of the Scores of Each Group Lying Between Its Median and the Median for the Neighboring Group

To Median of Group	PERCENTAGES WITH EQUIVALENTS, DISTRIBUTION OF		
	Three-Month Group	Six-Month Group	Nine-Month Group
Three-Month Group, Per Cent		41.22	46.95
P. E.		2.009	2.780
Six-Month Group, Per Cent	47.77		13.18
P. E.	2.978		.499
Nine-Month Group, Per Cent	49.27	15.55	
P. E.	3.621	.594	

between the six- and nine-month medians .449. Finally, by subtracting the six- to nine-month interval, .499 P. E., from the three- to nine-month interval, 2.780 P. E., we get 2.281 P. E., which is the other measure of the three- to six-month interval as determined by the nine-month distribution. This is known as the 'upper indirect' measure.

All the essential data for the computation of both the direct and indirect measures of the intergroup intervals are given in more concise form in Table XV.

For the final calculations of the distances between the medians of the three different groups by the 'overlapping method', double weight has been given to all direct measures and single weight to all indirect measures. These results are given in tabular form in Table XVI.

TABLE XVI

Determination of Intergroup Intervals From Overlapping of Distributions in Equation and Formula Test

Determination	Three- to Six-Month Interval	Six- to Nine-Month Interval
Lower Indirect		.643
Lower Direct	2.978	.594
Lower Direct	2.978	.594
Upper Direct	2.009	.499
Upper Direct	2.009	.499
Upper Indirect	2.281	
Total	12.255	2.829
Average	2.463	.566

Values for the distances between the group medians have now been worked out in three different ways. Each method possesses some merit; but the writer feels that the 'problem method' is perhaps the least reliable because it is subject to such extreme variations and that, all facts considered, the 'overlapping method' is the most consistent. It has, therefore, seemed wisest to attach double weight to the results obtained by the 'overlapping method' and single weight to each value obtained by the other two methods. The results of these calculations together with the final average are given in Table XVII. These averages will be used in the further development of the Equation and Formula Scale.

V. ESTABLISHING THE ZERO POINT

One thing more remains to be done before we can locate the exercises of the equation and formula test upon a linear scale. So far, we have merely located each exercise a specified number of units above or below the median performance of a group. We wish to know, ultimately, exactly how many units of difficulty each exercise represents and how many times as difficult one exercise is than another. To do this, it is necessary to locate each exercise with reference to a common zero point.

Zero points may be absolute or they may be arbitrary. Since it would be difficult to find exercises so easy that to be unable to solve them would signify that the student had absolutely no algebraic ability, it will be necessary to choose arbitrary zero points for these scales. Zero points on these scales will, therefore, mean

TABLE XVII

Final Determinations of Intervals Between Successive Groups for Equation and Formula Test [9]

	Three- to Six-Month Interval	Six- to Nine-Month Interval
Problem Method	1.434	.447
Quartile Method	2.671	.592
Overlapping Method	2.463	.566
Overlapping Method	2.463	.566
Average Used as Final Measurement	2.258	.543

simply the absolute inability to solve correctly any printed exercise which is as difficult as the easiest one used in these tests. The zero point, then, as used in these scales does not mean just not any algebraic ability at all, but it means simply the zero point for ability to solve the printed exercises as given in these tests. It is entirely possible that if the exercises had been presented orally to the students of the first, or three-month group, they would have solved more of them correctly. Some of the students who were unable to solve a single exercise, as the test was presented to them,

[9] Intergroup intervals for each of the other tests were determined by the same methods. These were found to be as follows:

	Three- to Six-Month Interval	Six- to Nine-Month Interval
Addition and Subtraction	1.387	.466
Multiplication and Division	1.685	.852
Problems	.689	.526
Graphs (Four and one-half to Nine-Month Interval)	2.366	

would in all probability have solved at least one of them if the test had been presented orally. Furthermore, the scales were developed independently of each other, therefore, each has its own zero point. We cannot say that the zero point on one scale is equal to the zero point on another, nor can we say that an exercise which has a value of one on one scale is of the same difficulty as an exercise that has the same value on another scale, because it is obvious that we cannot treat values as being equal which have been developed from different zero points.

The zero point for the Equation and Formula Scale was determined by the following methods. Table VI shows that 11 of the 689 students of the first group made a score of zero, which means that they received no score at all. Between these students and the median of the group (7.791) there were 333.5 students or 48.40 per cent of the whole group. Since this represents a deviation from the median or fifty per cent, it can be turned directly into a P. E. distance. Reference to Table X shows that a deviation of 48.40 per cent from the median of a normal distribution represents a distance of 3.182 P. E. According to this method, therefore, the median of the three-month group is just 3.182 P. E. above no score at all.

Another determination for the location of the zero point may be similarly obtained from the distribution of the six-month group. Table VI shows that only one student failed to make a score. There are, therefore, 372 students or 49.87 per cent of the whole group between the median achievement and those students who could not get a single exercise. This method would locate the six-month median 4.5 P. E. above zero. But we have just determined in the previous section (Table XVII) that the six-month median was 2.258 P. E. above the three-month median. Subtracting 2.258 P. E. from 4.5 P. E. we get 2.242 P. E., which is another value, though a somewhat more indirect one, of the distance the three-month median is located above the zero point.

Again, the median score of the three-month group as shown by Table VI is 7.791 and the quartile is 1.849. Assuming that the quartile is equal to the P. E., as it is in a normal surface of distribution, then by dividing the median achievement by the quartile we get 4.214 P. E., which is a third measure of the distance between our zero point and the median of the three-month group.

The Equation and Formula Scale 65

In the same way, by dividing the median achievement of the six-month group, which is 14.299, by the quartile for that group, which is 3.025, we get 4.727 P. E., the distance the six-month median is located above zero. Then subtracting the distance the six-month median is above the three-month median, 2.258 P. E., from the distance the six-month median is above zero, 4.727 P. E., we get 2.469 P. E., which is another indirect measure of the distance from zero to the three-month median.

We now have four determinations of the distance below our three-month median at which the zero point may be located. Undoubtedly those values based upon the results of the three-month group achievement are much more reliable, because the indirect measures—those based on the results of the six-month group—involve measurements taken at the utmost extremity of the curve (4.5 P. E.), and at this extreme range accident may unduly affect the results. It has, therefore, seemed best to give the indirect determinations single weight while the direct ones are given triple weight. The final determinations are as follows:

> From three-month distribution, 3.182
> From three-month distribution, 3.182
> From three-month distribution, 3.182
> From six-month distribution, 2.242
> From three-month achievement, 4.214
> From three-month achievement, 4.214
> From three-month achievement, 4.214
> From six-month achievement, 2.469
>
> *Final average*   *3.362*

Thus, we have established our arbitrary zero point for the equation and formula exercises[10] at 3.362 P. E. below the median of the three-month group.

With the median of the three-month group located at 3.362 P. E. above zero, and knowing the distance between the medians of each successive group, it is easy to determine the distance each group median is above the zero point. These distances are shown in Table XVIII.

---

[10] The distances of the arbitrary zero point below the median of the three-month group for the other exercises and problems are:

> Addition and Subtraction   3.826
> Multiplication and Division   4.067
> Problems   4.219
> Graphs (below four and one-half month group)   3.304

66   First Year Algebra Scales

TABLE XVIII

Distances the Median of Each Group is Above Zero (1 = P. E.)

Group	Above Zero	Below Next Group
Three-Month Group	3.362	2.258
Six-Month Group	5.620	.543
Nine-Month Group	6.163	

Fig. 9 illustrates graphically upon a linear scale the relative position of each of the group medians with reference to each other and with reference to their common zero point.

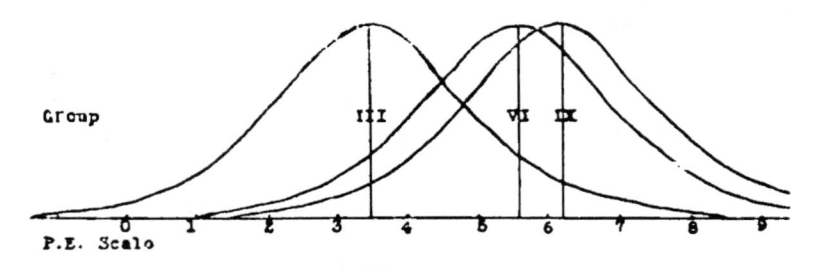

Figure 9
Relation to Each Other of the Three-, Six-, and Nine-Month Group Distributions in the Equation and Formula Test.

These values are taken from Table XVIII, and are, of course, based upon the assumption that achievement in the solution of the exercises is distributed in accordance with the normal surface of frequency.

VI. LOCATION OF EACH EXERCISE UPON A LINEAR SCALE

Before we are prepared to locate each exercise of the equation and formula test upon a linear scale and thus determine the general difficulty of each exercise, it is necessary to determine how far above the zero point each problem is located for each of the three groups. That is, we must refer each exercise to the common zero point. Table XI shows that exercise No. 1 is 2.688 P. E. below the median of the three-month group, but Table XVIII indicates that this median is 3.362 P. E. above the zero point. Therefore, the distance above zero at which this exercise is located for the three-month group is obtained by subtracting 2.688 P. E. from 3.362 P. E., which equals .674 P. E. Likewise, by subtracting 3.668 P. E. from 5.620 P. E., the distance the six-month median is located

TABLE XIX

Location Above Zero of Each Equation and Formula Exercise

Problem Number	Three-Month Group	Six-Month Group	Nine-Month Group
1	.674	1.952	2.004
2	2.011	3.074	3.515
3	1.713	2.688	3.119
4	2.366	3.443	3.941
5	2.689	3.825	4.254
6	2.783	3.527	3.997
7	3.451	4.480	4.931
8	3.209	4.573	4.936
9	3.742	5.705	5.432
10	3.425	4.868	4.998
11	5.221	4.751	4.915
12	4.759	5.351	5.640
13	5.643	4.910	5.373
14	6.267	5.433	5.444
15	5.025	6.484	6.234
16	7.005	5.351	5.407
17	7.005	5.992	6.107
18	6.662	6.124	6.033
19	7.005	6.826	6.714
20	7.637	6.906	6.828
21		7.636	7.537
22	7.300	7.076	7.182
23	7.445	7.256	7.149
24		9.015	8.287
25		9.411	8.985

above zero, we get 1.952 P. E., which is the location above zero of the same exercise as determined by the six-month group. In this way, the location above zero of each exercise, as determined by each of the three groups, was calculated. These results are recorded in Table XIX. As a further explanation of the method pursued in constructing this table, it should be noted that all negative values in Table XI indicate distances below, and that all positive values indicate distances above the median of a group. Wherever positive values are given in Table XI, it is, therefore, necessary to add this P. E. distance to the P. E. distance given in Table XVIII. For example, the location of exercise No. 15, as determined by the nine-month group, is 6.163 P. E.+.071 P. E. or 6.234 P. E. above zero.

We now have three values of the P. E. distance above zero at which each exercise may be located, one value as determined by the achievement of the three-month group, another as determined by the achievement of the six-month group, and a third as determined by the achievement of the nine-month group. Our final task is to determine the value of each exercise in general and to locate it upon one linear scale.

A study of Table XIX reveals the fact that there is a tendency for the easy exercises to be located higher up on the scale the more advanced the group; on the other hand, the difficult exercises are located farther up on the scale the less advanced the group. This we would naturally expect, since each group of students is likely to do its best, relatively speaking, upon those exercises which correspond most closely with the degree of advancement of the group. For this and other reasons, the most reliable measure of the difficulty of an exercise may be obtained from that distribution whose median of achievement is nearest to the actual position of the exercise on the scale; and, since our aim in these final calculations is to find that position which best represents the difficulty of each exercise, it will be necessary to weight our values accordingly. In the final determinations it was, therefore, decided to give quadruple weight to those values in Table XIX when the exercise is less than 1 P. E. distance from the median achievement of the group as seen in Table XI, triple weight if the exercise is more than 1 P. E. but less than 2 P. E. distance from the median achievement of the group, double weight if it is more than 2 P. E. but less than 3 P. E., single weight if it is more than 3 P. E. but less than 4 P. E.,[11] and not to consider it at all if it is beyond 4 P. E.

Fig. 10 shows the influence of the various parts of a distribution in determining the final value of an exercise according to the above system of weighting.

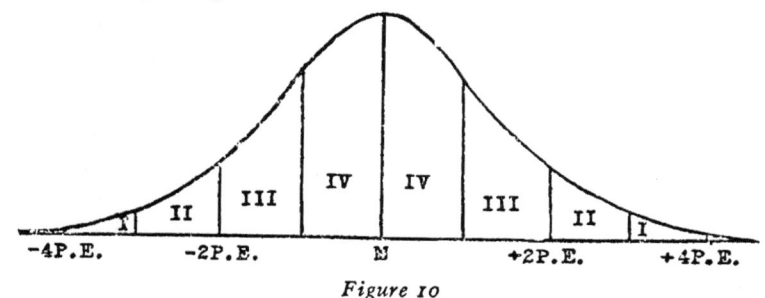

Figure 10

Showing the Influence of the Various Parts of a Distribution in Determining the General Value of an Exercise.

To illustrate the way this system of weighting works, let us calculate the general value of exercise No. 3.

	Value Table XIX	P. E. Deviation Table XI
Three-Month Determination	1.713	−1.649
Three-Month Determination	1.713	
Three-Month Determination	1.713	
Six-Month Determination	2.688	−2.932
Six-Month Determination	2.688	
Nine-Month Determination	3.119	−3.044
Average Determination	2.272	

In this way, the general value of each exercise was calculated. These values are given in Table XX. They are final values, and locate the distance of each exercise above an arbitrary zero point upon a linear scale.

[11] Inasmuch as our group medians were bunched so closely together, it was necessary to use the values at these extreme ranges more extensively than would otherwise have been necessary.

TABLE XX
Final Value of Equation and Formula Exercises

Rank	Problem Number [13]	Value
1	1	1.10
2	3	2.27
3	2	2.74
4	4	3.03
5	6	3.27
6	5	3.49
7	8	4.13
8	7	4.20
9	10	4.38
10	11	4.94
11	9	4.96
12	13	5.24
13	12	5.29
14	16	5.56
15	14	5.60
16	15	5.99
17	18	6.14
18	17	6.17
19	19	6.79
20	20	6.86
21	22	7.15
22	23	7.19
23	21	7.58
24	24	8.53
25	25	9.13

[13] Numbers given here indicate the position of the exercise in the test series.

IV
TABLES OF CRUDE DATA FROM WHICH THE OTHER SCALES WERE DERIVED
TABLE XXI
Distribution According to the Number of Addition and Subtraction Exercises Solved

	Three-Month Group	Six-Month Group	Nine-Month Group
0	7		
1	7		
2	10		3
3	14	4	5
4	29	1	9
5	44	11	5
6	73	19	21
7	84	30	28
8	98	36	35
9	117	54	50
10	105	62	86
11	111	95	146
12	132	114	155
13	41	67	135
14	19	56	104
15	7	52	111
16	2	37	124
17	3	44	117
18	3	34	88
19		36	71
20		30	54
21		26	40
22		15	28
23		7	25
24		2	4
Number Tested	906	833	1447
Median	9.744	12.908	14.409
25 Percentile	7.506	10.843	11.799
75 Percentile	11.824	16.615	17.558
Quartile	2.159	2.886	2.878

TABLE XXII

Number in Each Group That Solved Each Exercise in Addition and Subtraction Test Correctly

Problem Number	Three-Month Group	Six-Month Group	Nine-Month Group
1	867	819	1425
2	884	825	1433
3	833	812	1418
4	846	771	1377
5	653	735	1316
6	674	712	1259
7	679	765	1344
8	560	659	1202
9	637	680	1246
10	471	603	1105
11	393	574	1083
12	360	524	926
13	184	454	927
14	56	384	791
15	8	185	369
16	23	327	670
17	18	341	681
18	10	212	497
19	5	174	402
20	4	150	292
21	2	131	270
22		95	191
23		32	59
24		11	158
Number Tested	906	833	1447

TABLE XXIII

Location Above Zero of Each Addition and Subtraction Exercise

Problem Number	Three-Month Group	Six-Month Group	Nine-Month Group	Final Value
1	1.282	2.060	2.032	1.66
2	.902	1.741	1.777	1.18
3	1.752	2.308	2.200	2.06
4	1.592	3.079	2.791	2.46
5	2.957	3.456	3.265	3.19
6	2.854	3.644	3.581	3.30
7	2.830	3.149	3.067	2.97
8	3.381	4.012	3.823	3.70
9	3.036	3.878	3.635	3.46
10	3.752	4.331	4.177	4.08
11	4.072	4.482	4.248	4.27
12	4.213	4.724	4.713	4.55
13	5.058	5.045	4.709	4.92
14	6.107	5.358	5.069	5.39
15	7.332	6.348	6.221	6.41
16	6.731	5.620	5.382	5.74
17	6.870	5.554	5.352	5.61
18	7.221	6.095	5.844	6.11
19	7.646	6.414	6.117	6.42
20	7.764	6.570	6.482	6.70
21	8.101	6.706	6.562	6.63
22		7.001	6.900	6.95
23	8.426	7.927	8.258	8.09
24		8.513	7.506	7.76

TABLE XXIV

Distribution According to the Number of Multiplication and Division Exercises Solved

	Three-Month Group	Six-Month Group	Nine-Month Group
	3		
	5		
-	6		
3	28		
4	21	3	-
5	33	3	7
6	49	11	1
7	69	19	12
8	88	26	19
9	108	42	33
10	81	54	47
11	84	70	58
12	56	90	99
13	36	84	111
14	33	86	147
15	19	89	162
16	9	63	192
17	8	56	207
18	4	49	136
19		43	118
20		14	101
21		6	37
22		2	10
23			6
Number Tested	740	811	1505
Median	9.630	14.029	16.284
25 Percentile	7.579	11.625	13.885
75 Percentile	11.762	16.480	18.233
Quartile	2.091	2.427	2.174

TABLE XXV

Number in Each Group That Solved Each Exercise in Multiplication or Division Test Correctly

Problem Number	Three-Month Group	Six-Month Group	Nine-Month Group
1	724	804	1499
2	705	796	1486
3	662	784	1456
4	597	771	1452
5	674	771	1437
6	494	716	1385
7	535	710	1343
8	409	624	1210
9	503	667	1318
10	240	467	955
11	311	521	1053
12	264	535	1272
13	203	546	1145
14	99	497	1257
15	65	410	1167
16	136	377	937
17	72	200	497
18	28	250	727
19	46	179	529
20	5	112	361
21		12	118
22	2	166	592
23		5	56
Number Tested	740	811	1505

TABLE XXVI
Location Above Zero of Each Multiplication and Division Exercise

Problem Number	Three-Month Group	Six-Month Group	Nine-Month Group	Final Value
1	1.069	2.227	2.666	1.45
2	1.584	2.675	3.304	2.29
3	2.208	3.026	3.878	2.92
4	2.781	3.299	3.918	3.26
5	3.070	3.299	4.090	3.42
6	3.427	3.987	4.521	3.86
7	3.189	4.046	4.769	3.92
8	3.869	4.661	5.335	4.54
9	3.374	4.384	4.791	4.10
10	4.744	5.468	6.096	5.44
11	4.366	5.213	5.831	5.14
12	4.610	5.136	5.098	4.93
13	4.958	5.087	5.552	5.16
14	5.710	5.326	5.160	5.39
15	6.074	5.730	5.484	5.72
16	5.402	5.882	6.143	5.85
17	5.993	6.766	7.256	6.73
18	6.698	6.496	6.667	6.60
19	6.348	6.892	7.171	6.89
20	7.710	7.368	7.651	7.54
21		9.000	8.707	8.80
22	8.150	6.974	7.003	6.99
23		9.469	9.248	9.32

TABLE XXVII

Distribution According to the Number of Problems in Problem Test Solved

	Three-Month Group	Six-Month Group	Nine-Month Group
0	1		
1	19	7	6
2	25	26	18
3	70	47	69
4	154	96	122
5	130	118	161
6	115	126	174
7	67	97	194
8	39	80	182
9	16	51	139
10	10	39	96
11	1	20	74
12		8	33
13		2	30
14			7
Number Tested	648	717	1305
Median	5.423	6.512	7.528
25 Percentile	4.305	5.027	5.691
75 Percentile	6.765	8.259	9.379
Quartile	1.230	1.616	1.944

TABLE XXVIII
Number in Each Group That Solved Each Problem in the Problem Test Correctly

Problem Number	Three-Month Group	Six-Month Group	Nine-Month Group
1	640	712	1301
2	531	624	1165
3	579	682	1226
4	530	597	1144
5	346	505	949
6	215	304	728
7	223	326	666
8	101	206	551
9	40	142	431
10	57	121	323
11	6	126	400
12	10	83	252
13	2	9	43
14	2	16	125
Number Tested	648	717	1305

TABLE XXIX
Location Above Zero of Each Problem of the Problem Test

Problem Number	Three-Month Group	Six-Month Group	Nine-Month Group	Final Value
1	.892	1.245	1.376	1.07
2	2.868	3.218	3.401	3.16
3	2.376	2.435	3.151	2.61
4	2.873	3.456	3.724	3.35
5	4.092	4.093	4.549	4.24
6	4.863	5.172	5.228	5.09
7	4.815	5.056	5.407	5.09
8	5.718	5.722	5.736	5.72
9	6.500	6.147	6.096	6.20
10	6.226	6.309	6.458	6.36
11	7.790	6.268	6.192	6.42
12	7.438	6.660	6.730	6.80
13	8.302	8.234	8.170	8.19
14	8.302	7.874	7.388	7.58

TABLE XXX

Distribution According to the Number of Exercises in the Graph Test Solved

	Four and One-half Month Group	Nine-Month Group
	8	2
	27	5
2	92	10
3	165	30
4	102	73
5	47	113
6	28	122
7	15	140
8	5	129
9	3	102
10	3	38
11		3
Number Tested	495	767
Median	3.730	7.204
25 Percentile	2.965	5.635
75 Percentile	4.777	8.622
Quartile	.906	1.494

TABLE XXXI

Number in Each Group That Solved Each Exercise in the Graph Test Correctly

Problem Number	Four and One-half Month Group	Nine-Month Group
1	473	755
2	448	743
3	384	727
4	240	674
5	50	211
6	165	520
7	91	445
8	10	254
9	2	434
10		214
11		47
Number Tested	495	767

TABLE XXXII

Location Above Zero of Each Exercise in the Graph Test

Problem Number	Four and One-half Month Group	Nine-Month Group	Final Value
1	.882	2.473	1.41
2	1.360	2.903	1.98
3	2.179	3.259	2.61
4	3.360	3.935	3.61
5	5.196	6.556	5.97
6	3.944	4.977	4.46
7	4.639	5.371	5.06
8	6.348	6.318	6.32
9	7.242	5.424	5.79
10	7.242	6.539	6.68
11	7.242	7.963	7.72

APPENDIX

ADMINISTRATION AND USE OF FIRST YEAR ALGEBRA SCALES

Standard tests used in measuring schoolroom products must be as simple as possible, and their use must be thoroughly understood by the teachers who are expected to administer them. For the convenience of these teachers all suggestions concerning the use of the algebra scales are brought together here in concise form. Little new material is added, and what follows is largely a rearrangement of what has preceded, with possibly an added emphasis upon certain essential features.

To test a class in first year algebra all that is necessary is to procure a set of the scales, follow the detailed instructions in administering and in scoring them, and finally determine the median class score by the method as shown in this section. This median score can then be compared with the tentative standard scores given in this monograph.

TYPES OF SCALES

Five different scales are available. These are:
1. Addition and Subtraction Scale.
2. Multiplication and Division Scale.
3. Equation and Formula Scale.
4. Graph Scale.
5. Problem Scale.

Two of these, it will be seen, are designed to test the achievement of students in the fundamental operations, two of them test the ability of students in handling the instruments of quantitative thinking, and the last is composed of typical problems in first year algebra.

Two series of each scale are offered—Series A and Series B. Series B is the long one and contains from eleven to twenty-five exercises in each scale. Series A is shorter and contains from eight to twelve exercises in each scale. It is composed of exercises taken from Series B. It covers just as wide range of difficulty and has the added advantage of having the intervals between successive exercises or problems approximately equal.

SELECTION OF SCALES TO BE USED

The writer feels that Series A will be found to be on the whole more satisfactory than Series B, especially where the time available for testing purposes is limited. This is particularly true if the purpose of the test is primarily the determination of degrees of attainment. If, however, the purpose of the test is mainly diagnostic, that is, to discover difficulties which students are encountering, Series B should be used because it contains a more complete and richer variety of exercises.

If only one scale can be used, the writer would recommend the Equation and Formula Scale, because it is more comprehensive and so tests a much wider function. At least two scales should be used, however, and the writer feels that the Problem Scale should come second in importance.

WHEN TO GIVE THE SCALES

The comparative standards derived from this study and published in this monograph are for three-, six-, and nine-month intervals. It is, therefore, much more satisfactory for comparative purposes to submit these scales to algebra classes immediately after they have studied algebra for three, for six, or for nine months.

DIRECTIONS FOR GIVING SCALES

1. *Preliminary.* While the test sheets are being distributed, ask the students to wait for full directions before they do anything at all with the tests. When all are ready, say to the class, "Now write your name in the first blank space. In the next blank space to your right tell how old you are. Give this in full years to your nearest birthday," etc.

2. *Instructions.* After the sheets are properly headed, say to the class, "The exercises on this sheet are in addition and subtraction, collection of terms (multiplication and division, on the equation and formula, graphs, problems in algebra). Work directly on these sheets. Take the exercises in the order in which they are given. Work as many as you possibly can and be sure you get them right. If you come to one you cannot do, leave it out and pass on to the next."

Appendix 83

In the Multiplication and Division test, call attention to the fact that "all answers must be reduced to their *simplest* forms."

For the Graph Scale, let the students use rulers.

For the Problem Scale add: "In all the problems which call for the equation, *e. g.*, No. 4, simply state the equation which will solve the problem. That is, if you were given this problem: A coat and hat cost $30. The coat cost 5 times as much as the hat. Find the cost of each. The equation would be $x + 5x = \$30$." (Then write this equation on the board.)

NOTE. Students may be provided with scratch paper for their own use. It has been found to be most satisfactory to pass down the aisles and give each pupil a sheet of scratch paper a few minutes after the test has begun. They will find it most convenient, however, to work directly on the question sheets. For all but the problem test it is desirable to have as much of the work as possible on these sheets.

TIME LIMITS

The time limits to be observed in giving the tests are as follows: For each test in Series A allow *twenty* minutes, except the Problem and the Graph tests, for each of which *twenty-five* minutes should be allowed. For each test in Series B, allow *forty* minutes.

A warning, stating the amount of time left, should be given three minutes in advance for the tests of Series A, and five minutes in advance for those of Series B.

SCORING RESULTS

Problems are to be marked either right or wrong. All answers which may be accepted as correct are given in Tables I and II of this monograph.[1] After the scoring is completed, it will be found convenient to tabulate the results of a class as shown in Fig. 1.

DIRECTIONS FOR DETERMINING CLASS SCORE

The median class score is used in connection with these scales as the most satisfactory measurement of the achievement of a class. This median score represents the number of problems solved correctly by just fifty per cent of a class. That is, there are just as many students in a class who solve a larger number as there are students who solve a smaller number of problems. It is very readily computed; and, because of the fact that an error of one-half point in its computation becomes quite significant when comparison is to

[1] See pp. 30-36.

be made with the standard median scores given for the various scales of this study, a few sample class distributions are given below and their median class scores computed.

TABLE I

Sample Distribution of Scores made by Four Different Classes in Equation and Formula Test

Number of Problems Solved	CLASSES			
	I	II	III	IV
0				
1	1		2	
2	2	1	1	1
3	2		2	3
4	1		1	2
5	4	1	2	3
6	6	5	3	2
7	3	4	1	
8	3	3	1	3
9	2	2	2	4
10	1	1	1	2
11	1	1		
12				
13				
14		1		
15				
Total	26	19	16	22

According to this table, there were twenty-six students in Class I, nineteen students in Class II, sixteen students in Class III, and twenty-two students in Class IV. In Class I one student solved one problem correctly, two solved two problems correctly, two solved three correctly, etc. To find the median score of this class, it is necessary to find the point in the distribution of the class where there are just as many students who solved a greater number of problems as there are students who solved a smaller number. Since there are twenty-six students in the class, this point is obviously midway between the scores made by the thirteenth and

Appendix 85

fourteenth students, counting down in the distribution. That is, to include the thirteenth individual with the poorer group, it is necessary to count three of the six students who solved six problems; and, since it is assumed that the individuals are distributed evenly through a step at equal distances from one another, the median point is just one-half of the distance through this step, from six to seven. Therefore, the median score of this class is 6.5 problems solved correctly.

In Class II there are nineteen students. Thus, there are 9.5 individuals both above and below the exact median point in the distribution of this class. To include 9.5 individuals with the poorer group, it is necessary to count 2.5 of the 4 students who solved 7 problems. Hence, the median point is just $\frac{2.5}{4}$ of the distance through the 7th step, which makes the median score of this class 7.6 problems solved correctly. Class III illustrates another situation still. There are 16 students in the class, and in counting out the 8 individuals for the poorer group, we exactly take up all the cases in the 5th step. The fact to be noticed here is that the median point is raised clear through the 5th step. The median score for this class, therefore, is 6.0 problems correctly solved.

Class IV is here included to assist in the solution of another difficulty which is often encountered. There are 22 individuals in the class. Counting from the top of the distribution, the 11 cases for the poorer group take us, as seen above, entirely through step 6. Likewise, counting from below, to include 11 cases, we have to go clear through step 8. From this it appears that the median point could be located all the way from 7 to 8 in the class distribution. Since, however, any given distance on a scale is best represented by its middle point, the median score of this class should be 7.5 problems solved correctly.

TENTATIVE STANDARD SCORES [2]

As stated above, tentative standards of achievement have been established in all the scales for classes that have studied algebra

---

[2] To establish more accurate standards of attainment, the writer will be pleased to receive the results obtained by investigators who use any one of the scales proposed in this monograph. Such records should be sent to Teachers College Bureau of Publications.

for three, for six, and for nine months. These standards are given in Tables IV and V of this monograph.[3]

VALUE OF FIRST YEAR ALGEBRA SCALES

These scales may be used by teachers for three distinct and very useful purposes. They may be used (a) to indicate attainment, (b) to measure progress, and (c) to diagnose difficulties.

Scales which increase in difficulty by approximately equal steps furnish a most reliable objective means for determining the actual achievement of a student or a class of students. Any one of the scales may be used for this purpose, though the Equation and Formula Scale is perhaps to be preferred, since, as previously stated, it is a more comprehensive test. It is well to keep in mind also, in this connection that a low median class score is not always, nor even quite generally, due to poor instruction. Any one or a combination of several causes may be operating to keep a class score down. It is the duty of the teacher, however, to study these causes and to learn which ones are affecting the efficiency of the instruction, in order that proper remedial measures may be applied. This it is possible to accomplish with a much greater degree of certainty when the teacher knows the actual standard of achievement a class has attained. Such knowledge furnishes the teacher with a fact basis upon which to proceed and a motive with which to operate.

The extent of progress made by a class can be quite scientifically measured by submitting the same scale to a class at intervals of about three months. Teachers should be cautioned very specifically, however, not to do any drill work upon the exercises or problems appearing in the scales. If it is feared that some of the practice effect may survive, it is suggested that another scale in the same series, or the same scale in a different series, be used for the second test. The most desirable method of measuring progress, very naturally, would be to have another parallel series of scales similar and equal in difficulty to those of Series A, and it is to be hoped that such a series will soon be constructed.

For diagnostic purposes the scales of Series B have been found to be more serviceable. They offer a richer variety of exercises and, therefore, a greater number of type processes. Hence, a more com-

[3] See p. 41.

plete analysis of the mistakes made by students, and the difficulties they encounter, is made possible.

Finally, it must be stated emphatically that these are primarily power tests and as such should never be used for purposes of drill. Furthermore, with the time limits as now fixed, they are speed tests to a limited extent only. If a pure speed test is desired, the Standard Tests[4] devised by Dr. H. O. Rugg could be used to advantage. These would be particularly useful in determining whether a class has had sufficient drill upon the fundamentals.

[4] *School Review,* October, 1917, 25:546.